T0297953

Advances in Process Systems Engineering – Vol. 4

COMPUTATION OF MATHEMATICAL MODELS FOR COMPLEX INDUSTRIAL PROCESSES

Advances in Process Systems Engineering

Series Editor: Gade Pandu Rangaiah
(National University of Singapore)

Vol. 1: Multi-Objective Optimization:
Techniques and Applications in Chemical Engineering
ed: Gade Pandu Rangaiah

Vol. 2: Stochastic Global Optimization:
Techniques and Applications in Chemical Engineering
ed: Gade Pandu Rangaiah

Vol. 3: Recent Advances in Sustainable Process Design and Optimization
eds: D. C. Y. Foo, M. M. El-Halwagi and R. R. Tan

Vol. 4: Computation of Mathematical Models for Complex Industrial Processes
by Yu-Chu Tian, Tonghua Zhang, Hongmei Yao and Moses O. Tadé

Advances in Process Systems Engineering – Vol. 4

COMPUTATION OF MATHEMATICAL MODELS FOR COMPLEX INDUSTRIAL PROCESSES

Yu-Chu Tian
Queensland University of Technology, Australia

Tonghua Zhang
Hongmei Yao
Moses O. Tadé
Curtin University of Technology, Australia

World Scientific

NEW JERSEY • LONDON • SINGAPORE • BEIJING • SHANGHAI • HONG KONG • TAIPEI • CHENNAI

Published by

World Scientific Publishing Co. Pte. Ltd.
5 Toh Tuck Link, Singapore 596224
USA office: 27 Warren Street, Suite 401-402, Hackensack, NJ 07601
UK office: 57 Shelton Street, Covent Garden, London WC2H 9HE

British Library Cataloguing-in-Publication Data
A catalogue record for this book is available from the British Library.

Advances in Process Systems Engineering — Vol. 4
COMPUTATION OF MATHEMATICAL MODELS FOR COMPLEX INDUSTRIAL PROCESSES

ISBN 978-981-4360-93-7

Printed in Singapore

Preface

There have been quite a few books on numerical computing and its engineering applications; and the number of such books is still increasing. This book sets itself apart from these books in that it uses process engineering examples, but will nevertheless be useful to other engineers and scientists. Spread throughout the book are a few comprehensive and representative case studies to process engineering problems in Chapters 3 to 6, as well as some fundamental investigations in Chapter 2 and comparative studies in Chapter 7.

While the main focus of this book is on the computation aspect of complex industrial processes, attention is also paid to process modelling. This is due to the fact that process modelling largely relies on where and how the process models will be used. For the same process, different types of models may be built for different applications of the models. For example, a model for process control design may be quite different from that for process dynamics optimization; the former requires on-line implementation while the latter can be computed off-line. Therefore, the model computation should be investigated with the model application in mind.

One of our goals of this book is to combine the fundamental concepts of engineering computing of complex process models with practical applications. Although technologies and applications change relatively rapidly, the fundamental concepts evolve much more slowly and form the foundation from which new technologies and applications can be developed, understood, and evaluated.

Another goal of this book is to provide readers with step-by-step procedures for conducting industrial process computing. Various computing methods have been explained and demonstrated through comprehensive case studies of carefully-selected real industrial processes with industrial significance; and all these case studies and procedures have been verified in real industries, experimental rigs, and/or simulation environments. This will provide the readers with some useful frameworks for applications of engineering computing in fundamental research problems and practical development scenarios. Without the need of major changes, many of the case studies and procedures presented in this book are directly applicable to other industrial processes.

Our research and development on modelling and model computation of complex industrial processes have been reported in many publications. The third goal of this book is to summarize our recent achievements in this area from our widely dispersed publications.

This book is mainly intended for academic researchers and industrial practitioners who work in control and optimization oriented process modelling and model computation. It is designed to follow the logic flow of numerical computing fundamentals, various numerical computing methods with case studies, and comparative studies of these computing methods through case studies. The readers with a fair background of applied mathematics and process systems engineering, e.g., third-year undergraduate students and postgraduate students, should be able to read this book comfortably without major difficulties.

The materials included in this book are emanated from the synergistic collaborations between two research groups: the group led by Professor Yu-Chu Tian at Queensland University of Technology, and the group led by Professor Moses O. Tadé at Curtin University of Technology. They reflect the achievements of all four authors in fundamental investigations and practical developments in complex industrial process computing.

The team of the four authors of this book are well blended with complementary skills and a long history of collaborations for research and development in computation of mathematical models for complex industrial processes. It includes a computer scientist (Y.-C. Tian), a mathematician (T. Zhang), and two chemical engineers (H. Yao and M. O. Tadé). Working at three different universities at the moment, the team members are leading research and development on different aspects of computational theory for complex industrial processes and other systems.

Acknowledgements

The work presented in this book was supported in part by various funding agencies and organizations, which the authors would like to acknowledge, including the Australian Research Council (ARC) under the Discovery Project Scheme (grant number DP0559111 to Tadé and Tian, grant number DP0770420 to Zhang), the Australian Government's Department of Innovation, Industry, Science and Research (DIISR) under the International Science Linkage (ISL) Project Scheme (grant number CH070083 to Tian), Baoshan Iron and Steel Corporation (BISC) under its Strategic Research Project Scheme (a grant in 1992 to Tian and his colleagues), and the Postdoc Foundation of China under its Research Project Scheme (a grant in 1994 to Tian).

The authors would also like to thank the editor of the book series, Professor Gade Pandu Rangaiah, for his encouragement, creative discussions, and careful editing, without which the publication of this book would have not been possible.

Last but not the least, special thanks go to Mr Steven Patt, Senior Editor, World Scientific Publishing, for his professional management of the whole process of the publication of this book. It has been enjoyable to work with him and the Publisher.

About the authors

Dr Yu-Chu Tian is a Professor of Computer Science and the Head of the Discipline of Networks and Communications, School of Electrical Engineering and Computer Science, Queensland University of Technology, Brisbane, Australia. Email: y.tian@qut.edu.au.

Dr Tonghua Zhang is a Senior Lecturer in Applied Mathematics, Discipline of Mathematics, Swinburne University of Technology, Melbourne, Australia. Email: tonghuazhang@swin.edu.au

Dr Hongmei Yao is a Senior Research Fellow in Chemical Engineering, Department of Chemical Engineering, Curtin University of Technology, Perth, Australia. Email: H.Yao@exchange.curtin.edu.au.

Dr Moses O. Tadé is a Professor of Process Systems Engineering and the Dean of Engineering, Curtin University of Technology, Perth, Australia. Email: M.O.Tade@curtin.edu.au

Contents

List of Figures

List of Tables

List of Tables

Chapter 1

Introduction

This book focuses on computation of mathematical models for complex industrial processes. As the computation serves a specific purpose in a real application, it is tightly related to process modelling, from which the mathematical models are developed for the purpose. For an industrial process, different types of mathematical models may be required for different applications of the models. For example, models for real-time process control may be different from those for process optimization. Therefore, process modelling will also be discussed in this book for better understanding of model computation methods and technologies.

While computation of mathematical models is a very broad topic, this book will investigate computational problems from the practical application perspective. Nevertheless, special attention will also be paid to fundamental concepts and theory of mathematical computation in problem solving for real industrial processes. This is based on our understanding that fundamental concepts and theory evolve much more slowly than technologies and applications and form the foundation from which new technologies and applications can be developed.

In the development of the content of this book, step-by-step procedures will be provided for practical industrial process modelling and model computation. Comprehensive case studies will also be demonstrated through carefully selected and experimentally verified complex industrial process examples. These procedures and case study examples form some useful problem solving frameworks, which the readers may follow in real industrial process applications.

1.1 Background

What is *computation of mathematical models*?

In this book, computation of mathematical models refers to a set of techniques with which the mathematical models formulated for a specific problem can be solved through arithmetic and logical operations. This is a traditional understanding of the computation of mathematical models and gives the focus of the computational problem onto the development and theoretical justification of various computing methods. Many such traditional numerical analysis books have been published, which were written in this way with detailed theoretical derivations.

The computation of mathematical models can also be understood as a process of deriving a solution with a computer or multiple computers from a given set of mathematical models. With this understanding, the focus of the investigations would be on problem solving, i.e., on the choice and application of numerical methods for deriving a satisfactory solution. Theoretical derivations and rigorous analysis of the computation may be largely ignored or not fully discussed. For a specific computational problem of mathematical models, techniques and algorithms could be recommended for this type of models from previous experience of successful applications. To a large extent, examples of such successful applications form a foundation of useful frameworks with which many practical problems could be solved straightaway without the need to carefully consider related mathematical theory.

Computation of mathematical models is discussed in this book in a different way from the above two typical understandings in both intend and content. While intending to emphasis more on solving practical problems with appropriate frameworks and techniques which are well verified and validated through successful applications, we still discuss necessary mathematical background and theoretical analysis. The content of the book is designed so that we could cover the three major types of effective methods and algorithms for computation of mathematical models of complex industrial processes: finite-difference methods, wavelet-based methods and high-resolution methods.

Computation of mathematical models serves a specific purpose in a real application. This book addresses more on model computation for online optimization and real-time control. Therefore, computational methods which we believe are not quite suitable for this purpose will not be investigated in this book. One of such methods is the widely used and well developed finite element technique for process design and optimization. Another type of such methods is those for fluid dynamics analysis, which is a broad topic of research and development in process systems engineering.

As digital computers are used for solving the computational problems, the computation we are investigating is actually numerical computation. Numerical computation by digital computers is naturally conducted by methods of approximation. This means that approximation methods are critical to the computation of mathematical models. To justify the approximation methods, mathematical justification needs to be provided for a number of issues, such as accuracy and precision of the approximation methods, computational complexity, computational stability, and the speed of convergence. For a specific computational problem of an industrial process model, an important task is to choose a right approximation method for the numerical computation in order to get the solution within an acceptable period of time under the constraints of the available computing resources.

Depending on the scale and complexity of an industrial process model, numerical computation of the mathematical behaves with different performance in computing resource requirements and execution time. In some scenarios, it can be completed very quickly without demanding excessive computing resources. In many other cases, it may be computationally expensive with a high demand of computing resources and significant time consumption. With the rapid development of modern computer technologies, many computational tasks of industrial process models can be easily handled with existing numerical methods. Such computing tasks would not be comprehensively discussed in this book.

Without discussing easy tasks in process model computing, this book will address computational problems for complex industrial processes, which require significant computing effort and thus demand innovative numerical methods. The complex processes we will cover in this book include galvanizing processes, biological fermentation processes, crystallization processes, chemical reaction processes and simulated moving bed chromatographic separation processes. These processes are either typical unit operation processes in process industries or an integration of several unit operation processes. They are also widely deployed in real process industries, and are thus industrially significant.

We emphasize again that the computation of a mathematical model serves a specific purpose for an application. In this sense, numerical methods for computation of mathematical models should be chosen to meet the requirements for the purpose. This means that from a mathematical model of an industrial process, different computational methods may be required for different application purposes. This highlights the need of careful investigations into the application requirements and the performance of the chosen computational methods.

In order to fulfill the application requirements, process modelling needs to be investigated before the computational problems of the mathematical models are tackled. We consider process modelling as an integrated and significant step in the whole procedure of the computation of mathematical models for complex industrial processes. Therefore, it will be discussed in this book for each of the selected industrial process examples.

Process modelling itself is a broad topic of research and development in process systems engineering. A complete discussion of process modelling is beyond the scope of this book. In general, a process model can be established from one of the following three methods:

(1) from the first principal and theoretical development;
(2) from empirical analysis and experimental studies; and
(3) from a combination of the above two methods.

In our observation, the majority of process models for real industrial applications are established through the third modelling method, i.e., through a combined application of theoretical development and empirical analysis. This is also the method we will adopt in this book for establishment of mathematical models of complex industrial processes. Techniques and approaches we will use for process modelling will be embedded into the discussions of our process modelling.

The results derived from numerical computation of mathematical models need to be carefully validated and verified before they are applied to real systems. This is because of two main reasons. The first reason is the approximation nature of numerical computing, which works in discrete-time domain to approximate system dynamics in continuous-time domain. Obviously, a discrete-time representation with finite number of time instances may not well approximate a continuous-time problem with infinite number of time instances. The second reason is the uncertainties in the mathematical models of the industrial process under investigation. Industrial processes in chemical industries are known to show significant uncertainties, which cannot be well captured in process modelling. A detailed discussion of this topic will not be carried out in this book. Interested readers are referred to process modelling books for comprehensive discussions on this topic.

1.2 Motivation

Why should we study computation of mathematical models for complex industrial processes?

Numerical computation of mathematical models is an integrated part of process optimization and real-time control in modern process industries. Optimization and control of industrial processes rely on various optimization strategies and control actions in real-time. These strategies an actions are typically designed from mathematical models of the processes. Therefore, numerical computation of mathematical models of the processes becomes an integrate part of process optimization and real-time control for many processes in modern process industries.

Most processes in modern process industries can be well optimized and controlled by using simple strategies such as proportional-integral-derivative (PID) algorithm. For these processes, the process dynamics are well understood and the implementation of numerical methods for process model computation is mature. Solutions to the process models can be obtained numerically with an acceptable consumption of computing resources and execution time. Thus, there is no need to design advanced and sophisticated computational methods for computation of the mathematical models. Therefore, this book does not address computational problems of such processes.

However, there are many complex industrial processes which behaves with complex behaviours or a large number of degrees of freedom. Nonlinear dynamics, multiplicity which shows co-existence of several steady states, bifurcation and a huge number of coupled model equations are just a few of many examples. To capture the dynamic features of such complex systems, effective methods are required in order to efficiently derive numerical solutions to the process models effectively and efficiently.

Also, modern process industries tend to increasingly integrate multiple processes into a single unit or a tightly coupled production line for improved system performance and/or reduced energy consumption. For example, integration of chemical reaction and distillation separation into a single unit leads to the design of compact reactive distillation columns. As another example, integration of multiple chemical reactors, each of which may exhibits complex spatiotemporal concentration patterns and other dynamics, leads to tighter system coupling, more delay variables, and more types of symmetry, inducing more complicated bifurcation and co-existence of multiple steady states. A further example is a continuous galvanizing production line, which will be discussed later in Chapter 4, is an integration of multiple and tightly coupled processes. Integration of multiple processes in modern process industries results in large-scale and complex process models, which require significant computing effort for numerical computation.

Investigations into efficient computation of mathematical models for complex industrial processes also greatly enhance our capability to tackle many problems that would otherwise be considered to be too complicated to handle. Typically, such problems may not be easily solved analytically and thus require effective and efficient numerical methods for model computation. Numerical computation makes it possible to deal with complex process dynamics and large-scale model problems with affordable computing resources and execution time.

Furthermore, a good understanding of model computation theory underlying various numerical methods enables appropriate use of software tools to solve computational problems for complex industrial processes. Nowadays, there are many commercial and open source software packages available for numerical computation of mathematical models. On the commercial side, typical examples are MATLAB, Maple, and Aspen Plus. Among open source software packages is SciLab. Knowledge of theoretic developments of computing methods is crucial for deriving useful numerical results and interpreting the results.

Finally, many computational problems of mathematical models cannot be directly solved using existing software packages. In this case, a good understanding of computation of mathematical models will facilitate design and development of computer programs to solve the problems. There are many occasions that we have to write our own programs for model computational problems under investigation.

With all the above mentioned reasons, we are well motivated to study computation of mathematical models for complex industrial processes. We are now ready to describe several specific and major problems we are going to address in computation of mathematical models for complex industrial processes. They include process modelling, model approximation, algorithm design and setup, and interpretation and verification of results. When discussing these problems, we will also identify some challenges around the problems.

It is worth mentioning that there are also many other issues in the broad area of process modelling and model computation. However, these issues are not the focus of this book dedicated to model computation for complex industrial processes.

1.3 Process Modelling

Process modelling is the first major problem to be addressed in this book for computation of mathematical models for a specific application. Depending how and where the process models to be established will be used, methods for process modelling may differ significantly. For example, in comparison with online process monitoring, real-time control typically requires much faster computation of process models and model-based control actions. In this sense, it is crucial to build process models with sufficient details of the process dynamics yet simple enough for efficient computation. Furthermore, a decision has to be made on whether the process models should be established through theoretical derivation or experimental investigations.

In modeling complex industrial processes, a significant challenge is to understand the process dynamics. This has not been an easy task. One reason is that some industrially important process variables, which are ideal control variables, are not measurable in real time with existing sensing technologies. An example is industrial crystallization processes, in which the quality of crystals as represented by crystal size and its distribution in the crystallizer cannot be directly measured in real time. Therefore, if a process model is established with consideration of these control variables, computational results from the mathematical model for the control variables cannot be experimentally verified in real time and thus are not reliable. How to charaterize these variables in process modelling to enable practical process optimization and real-time control becomes challenging.

For highly integrated industrial processes, there exist difficulties in capturing complicated process behaviours in a mathematical model that is sufficiently accurate to describe the process dynamics yet simple enough for practical computation. This is another reason why understanding process dynamics is challenging. A compromise needs to be found between the accuracy and simplicity of the mathematical models.

Furthermore, the dynamics of many complex industrial processes have not been well understood so far. For example, how will increased delay variables complicate the dynamics of bifurcation and co-existence of multiple steady states in integrated multiple chemical reactors is still not clear. This also makes the understanding of process dynamics challenging.

As a special topic in system modelling, system identification is also challenging, particularly for non-stationary dynamics with colour noise in process measurements. Quite a few books have been published in theory and applications of system identification techniques have been published. This book will employ some system identification techniques, but will not analyze system identification theory in detail in order not to distract readers from the main theme of this book.

1.4 Model Approximation

The second major problem to be highlighted in this book is how to approximate process models for numerical computation. Most mathematical models of industrial processes are naturally continuous in time and space. For numerical computation, they need to be represented in a form discrete in time and space through approximation. This will be discussed in detail in this book through well designed case study examples from practical industrial processes.

Many approximation methods have been developed each with its own features. For example, finite difference methods, finite element methods, wavelets-based methods and high resolution methods are just a few of many such methods. Each of these methods has a few variants. For some processes, all these methods and their variants many work well with good quality of the derived computing results. For other processes, one method may work better or even much better than others in terms of the consumption of computing resources and the quality of model solutions. Therefore, how to choose a right approximation method for a specific process model is challenging.

1.5 Algorithm Design and Setup

The third problem that will be investigated is algorithm design and setup. After an approximation representation of a process model is established, computing strategies and algorithms need to be designed and implemented. Meanwhile, values of many algorithm parameters have to be determined in order to practically implement the designed algorithms.

As in process model approximation, algorithm design also considers a compromise between the accuracy and resource demand of the computing. Understandably, a high accuracy of model computing requires higher computing effort. In many scenarios, an increase in computing accuracy leads to an exponential increase in computing resource demand. This becomes challenging for large-scale process models. However, limiting the consumption of computing resources including execution time may give the computing results that are inaccurate and thus do not meet the application requirements.

In determination of parameters in the approximated models and computing algorithms, not only the accuracy and resource demand of the computing need to be considered, but also the stability and convergence of the computing algorithms should be guaranteed to ensure the applicability of the computing algorithms. Typical model and algorithm parameters are time step and space step for the discrete-time models approximated from their continuous-time versions. Other parameters include many physical, chemical and other constants. Termination condition of the computing is also a parameter to be predetermined, which specifies at what accuracy or by what time the computing should be terminated with feasible computing results.

While using a single computer or processor platform for process model computation in practice, there are motivations to conduct model computation in multiple computers or a computer with multiple cores. Parallel and distributed computing is a computing technology to deal with large-scale computation or computational problems which require a large amount of data. Resource management and scheduling, job tracking, and communication and coordination among multiple computers are among many important issues in distributed computing. While distributed computing is an active area of research and development, it addresses more on pure computing techniques and less on process model computation. Therefore, this book will not specifically discuss this topic.

1.6 Interpretation of Verification of Computing Results

As the last problem to be addressed in this book, numerical results from model computation should be interpreted appropriately, and verified experimentally or in simulation. As numerical computation of a process model is conducted with approximation, the results from the computation may not reflect the real dynamics of the process under investigation. It is not an uncommon mistake that the computing results from mathematical models are directly adopted without testing and verification. Without testing and verification, the computing results become largely devalued, and the reliability and confidence of the computing results may also be questionable. Simulation, statistical analysis and experimental studies are widely used methods for testing and verification of computing results.

Simulation is a flexible and cheap technique for model verification. It can give the computing results easily and quickly for a wide range of operating conditions of the process models, enabling fast detection of abnormal events and dynamics. It can also be used to investigate the feasibility of process design, control design, and process optimization before a process is actually deployed.

Statistical analysis of the simulation or experimental studies is a collection of methods and techniques to process large amounts of measured data and to examine overall trends. It helps check how reliable the computing results are. Under a specific operating condition, statistical analysis of multiple runs of simulation or experimental studies gives the mean value, variance and other statistics. Conducting some statistical tests, e.g., t-test that is a statistical hypothesis test, will help find out at what confidence level the computing results are reliable.

Experimental investigations are a direct and convincing method for verification of numerical computing results of a process model. However, they are not always possible, particularly before an industrial process has not been actually deployed. Even if experimental studies are possible, they may be expensive to conduct due to the use of equipment, materials, energy, and personnel. Interruption of experimental studies to normal industrial production is also considered to be expensive in many cases. Therefore, experimental design is particularly important for experimental verification of model computation.

1.7 Book Outline

This book is divided into seven chapters. Each of these seven chapters addresses a specific topic with consideration of the major problems and challenges identified in the last few sections.

The book begins with this introductory chapter, which introduces the background and basic concepts of computation of mathematical models for complex industrial processes. The chapter also identifies major problems and challenges in the broad area of process modelling and model computation.

Chapter 2 introduces fundamental theory and techniques for computation of mathematical models for complex industrial processes. It covers ordinary differential equation (ODE) models and partial differential equation (PDE) models, solutions for different types of ODE and PDE models. Three main types of numerical techniques are investigated for solving process models: finite difference methods, wavelets-based methods and high resolution methods.

From Chapters 3 to 6, the finite difference methods, wavelets-based methods and high-resolutions methods are respectively discussed in detail. The discussions are carried out from process modelling to model computation with carefully selected practical examples. The examples include fermentation processes and galvanizing processes with finite difference methods, chemical reaction processes with wavelets-based methods, and column chromatographic separation processes with high-resolution methods, respectively.

Comprehensive comparative studies are conducted in Chapter 7 among the three main types of numerical computation methods discussed above. The complex simulated moving bed chromatographic (SMBC) process is selected as a representative process for the comparative studies.

Finally, Chapter 8 concludes the book.

Chapter 2

Fundamentals of Process Modelling and Model Computation

This chapter introduces fundamental theory and techniques for computation of mathematical models of complex industrial processes. It discusses, with examples, how to establish ordinary differential equation (ODE) models and partial differential equation (PDE) models from the first principles for industrial processes. Then, it investigates solutions for initial value ODE problem, boundary value ODE problem, and general PDE models, respectively. After that, four main types of computing techniques are briefly described for numerically solving process models: The Runge Kutta methods, finite difference methods, wavelets-based methods and high resolution methods.

As we have mentioned earlier in Section 1.1 of Chapter 1 (Introduction), this book addresses more on model computation for online optimization and real-time control. computational methods not quite suitable for this purpose will not be discussed in this book, such as finite element methods and those for comprehensive fluid dynamics analysis.

2.1 Building Mathematical Models

A mathematical model of a physical process is a set of coupled or independent equations that describe the dynamics of the process. It characterizes the states of the process dynamics by a number of process variables. Some of these process variables are independent, and others are dependent. Typical independent variables are time and space variables, while dependent variables characterize the state of the physical process. Examples of dependent variables are temperature, pressure, flow rate, and concentration of a component. In a very general form, dependent variables are represented as a function of independent variables

$$\begin{aligned}&\text{Dependent variable}\\&\quad = \mathit{function}(\text{independent variables, parameters, external force}). \quad (2.1)\end{aligned}$$

If the functional relationship in Equation (2.1) is known, we actually know the solutions of the dependent variables. The solutions can thus be directly adopted in various applications such as process optimization and control.

However, in industrial applications, such a direct functional relationship is not easy to establish. In most cases, dependent variables are described by a set of ODEs and/or PDEs, which may be coupled in different ways. If analytical solutions to these model equations, we will have to numerically solve the model equations.

The fundamental principles governing general physical systems are Newton's second law and conservation laws. They play an important role in building mathematical models for industrial processes. Conceptually, the conservation laws can be interpreted in process modelling as

$$\begin{aligned}&\text{Change or accumulation}\\&\quad = \text{Input or increase} - \text{Output or decrease}. \quad (2.2)\end{aligned}$$

This is the basic form of mathematical models built from the first principles for industrial processes. It will lead to various types of differential equation models.

Consider a cylinder tank of radius r to store incompressible fluid. We aim to describe the dynamics of the fluid level L in the tank. Because the volume V of the fluid in the tank is $V = \pi r^2 L$, the change in volume V at time t is $\frac{V}{dt} = \pi r^2 \frac{dL}{dt}$. It is the difference between the flow in rate in volume R_{in} and flow out rate in volume R_{out} at that time. This is mathematically described as

$$\pi r^2 \frac{dL}{dt} = R_{in}(t) - R_{out}(t). \quad (2.3)$$

If this is a single fluid level process, the flow in rate R_{in} can be an external force, which may not be regulated. The flow out rate R_{in} may be dependent on the value of the fluid level L in the tank, and the functional relationship between these two variables is generally a nonlinear function denoted as $f(\cdot)$. That is

$$R_{out}(t) = f(L(t)). \quad (2.4)$$

Substituting Equation (2.4) into Equation (2.3) gives an ODE model of L

$$\frac{dL}{dt} = \frac{1}{\pi r^2} R_{in}(t) - \frac{1}{\pi r^2} f(L(t)). \tag{2.5}$$

To derive a final version of the ODE model from Equation (2.5), an explicit expression is required for the nonlinear function $f(\cdot)$. This is obviously based on a good understanding of the underlying physical system.

In general, in order to establish mathematical models from the first principles for industrial processes, understanding the fundamental principles of the physical processes is important for process modelling. In industrial processes, mass momentum and energy transfer are the main physical phenomena, whose characteristic determine the state of the process variables such as temperature, flow fields and chemical concentrations. For chemical reaction processes, the reaction kinetics largely determine the dynamics of the chemical reactors. In biological processes such as industrial fermentation, the growth of biomass characterizes the dynamic behavours of the processes.

Typically, those physical, chemical and/or biological phenomena are described by a set of equations, such as algebraic equations, ODEs and PDEs. Therefore, as discussed previously, in order to understand the dynamical behavior of industrial processes, it is necessary to understand how to build process models of algebraic equations, ODEs and PDEs, and how to find suitable solutions for those equations.

For linear process models, which are described by linear equations, it is relatively easy to find analytical solutions. Thus, the computation of the mathematical models also becomes relatively easy because the solutions can be directly derived from analytical results. However, it will become difficult if the requirement for the computing time is very tight on embedded computing platforms with limited computing resources. Real-time control through embedded controllers is such an example. Furthermore, computation of large-scale models demands significant computing effort even though the models are linear.

In industrial applications, there are many process models described by nonlinear ODE and/or PDE equations. While analytical solutions may be derived for a specific nonlinear model, it is not realistic to find analytical solutions for general nonlinear equation models. Therefore, numerical solutions are required in practice to handle nonlinear model equations. Studies on effective and efficient techniques for numerical computation of process models is, and will still be, an active topic of research and development.

The following sections will introduce fundamental theory for model development and numerically solving ODE and PDE models. Examples will also be given for demonstration.

2.2 General ODE and PDE Models for Industrial Processes

All differential equations used in process modelling can be classified into two categories according to how independent variables appear in the equations: Ordinary Differential Equations (ODEs) and Partial Differential Equations (PDEs).

In a simple ODE equation, only one independent variable and one dependent variable. The following equation is a general form of an ODE with order of n

$$a_n \frac{d^{(n)}y(t)}{dt^n} + a_{n-1}\frac{d^{(n-1)}y(t)}{dt^{n-1}} + \cdots + a_1 \frac{dy(t)}{dt} + a_0 y = f(y(t), t), \quad (2.6)$$

where $t, y(t)$ are independent and dependent variables, respectively; t is time in particular; $a_0, a_1, \cdots, a_n$ are coefficients, which are constants in typical scenarios.

For a simple PDE, there are multiple independent variables but one dependent variable in a differential equation. A second-order PDE with independent time variable t, two additional independent variables x and y, and a dependent variable $u(x, y, t)$ is given below

$$\begin{aligned} &A\frac{\partial^2 u(x,y,t)}{\partial x^2} + B\frac{\partial^2 u(x,y,t)}{\partial x \partial y} + C\frac{\partial^2 u(x,y,t)}{\partial y^2} \\ &+ D\frac{\partial u(x,y,t)}{\partial x} + E\frac{\partial u(x,y,t)}{\partial y} + Fu(x,y,t) = G \end{aligned} \quad (2.7)$$

where A, B, C, D, G are constants in typical applications, and are determined by the physical properties of the process under investigation.

Most ODE and PDE models of industrial processes have similar forms to Eqs. (2.6) and (2.7) or a combination of both. Therefore, we will focus our discussions on these forms of ODE and PDE models in this book.

For industrial processes governed by ODEs, the equations can be first-order or higher-order. Since a higher-order ODE can be generally converted into a set of first-order ODEs, a set of first-order ODEs can be represented as a simple first-order ODE in vector form. Thus, we will focus our development of the numerical techniques on first-order systems, namely $n = 1$ in the rest of this book, unless otherwise specified explicitly.

To investigate solutions to ODE models, it is helpful to classify the ODE problems into **Initial Value Problems** and **Boundary Value Problems** according to the given conditions for solving the ODE models. An initial value problem is an ODE with a specified value, called the initial condition, of the dependent variable at a given point in the domain of the solution. In computation of mathematical models for industrial processes, initial value problem is quite popular, which specifies how, given initial conditions, the system state evolves with time. For example, for ODE model in Equation (2.6), an initial condition of y can be given at time t_0 as

$$y(t = 0) = y_0. \tag{2.8}$$

Eqs. (2.6) and (2.8) form an initial value problem of the ODE.

A boundary value problem is an ODE with a set of additional restrained values, called the boundary values, of the dependent variable at given points in the domain of the solution. For example, for ODE model in Equation (2.6), in addition to the condition in Equation (2.8), a further condition is given as

$$y(t = 10) = y_{10}. \tag{2.9}$$

Eqs. (2.6), (2.8) and (2.9) form a boundary value problem of the ODE.

In comparison with ODEs, PDEs are much more complicated due to the involvement of multiple independent variables in a single differential equation. According to the sign of $B^2 - 4AC$, the second-order partial differential equation given in Equation (2.7) is generally known as Elliptic type, Parabolic type or Hyperbolic type, respectively, as summarized in Table 2.1. Furthermore, in order to guarantee the PDE has a unique solution, both initial conditions and boundary conditions should be given.

Table 2.1 Classification of the PDE described in Eq. (2.7).

Sign of $B^2 - 4AC$	Classification
Negative	Elliptic type
Zero	Parabolic type
Positive	Hyperbolic type

2.3 Examples of ODE and PDE Process Models

This section gives two examples of process models. One example is in ODEs, and the model is established for an enzyme reaction process. The other example is two PDE models, which describe population balance and the Euler-Tricomi Equation, respectively, in process systems engineering.

2.3.1 *ODE Model for Enzyme Reaction*

Enzyme is commonly used in process industries. It is used in chemical plants when extremely specific catalysts are required. As described in [Murray (2002)], the mechanism for a single substrate enzyme catalyzed reaction can be schematically represented as

$$S + E \underset{k_{-1}}{\overset{k_1}{\rightleftharpoons}} SE \xrightarrow{k_2} P + E, \tag{2.10}$$

where k_i $(i = -1, 1, 2)$ are rates of reactions; S, E and P are substrate, enzyme catalyst and reaction product, respectively. Let s and c denote the concentrations of the reactants S and E, respectively. After some simplifications, we have a system model of ODEs

$$\begin{aligned} \frac{ds}{dt} &= -k_1 e_0 s + (k_1 s + k_{-1})c, \\ \frac{dc}{dt} &= k_1 e_0 s - (k_1 s + k_{-1} + k_2)c, \end{aligned} \tag{2.11}$$

with initial conditions

$$s(0) = s_0, c(0) = 0. \tag{2.12}$$

2.3.2 *PDE Model for Population Balance*

A PDE model is established for an industrial process where most of the problems to be addressed involve mass and energy balance, which is also referred to as Population Balance Equation (PBE). The governing equation of the population balance is given as follows

$$\frac{\partial n}{\partial t} + \frac{\partial G n}{\partial v} = f(v, t), \tag{2.13}$$

where n stands for number density of the population, G represents growth rate, v is particle size, and t is time variable. For different processes, G could be a constant or a function of particle size v. The form of function $f(\cdot)$ might be different as well.

Equation (2.13) is a special case of the general PDE model in Equation (2.7) with $A = B = C = 0$. It is seen from Table 2.1 that the PDE model in Equation (2.13) is a parabolic type because $B^2 - 4AC = 0$.

2.3.3 *PDE Model for Transonic Flow*

An additional example of PDE models is the mathematical representation of transonic flow, the so-called Euler-Tricomi equation. The PDE model is given by

$$y\frac{\partial^2 w(x,y)}{\partial x^2} + \frac{\partial^2 w(x,y)}{\partial y^2} = 0, \tag{2.14}$$

where x and y are two independent variables, and w is a dependent variable. The PDE model in Equation (2.14 is also a special case of the general PDE model in Equation (2.7) with $A = y, B = 0, C = 1$. It is easy to verify that $B^2 - 4AC = -4y$, which is a function of y, one of the independent variables. Depending on the value of y, the PDE model in Equation (2.14) could be any of the three types indicated in Table 2.1. To be more specific, we have

$$\text{The PDE in Equation (2.14) is} \begin{cases} \text{Hyperbolic, if} & y < 0, \\ \text{Parabolic, if} & y = 0, \\ \text{Elliptic, if} & y > 0. \end{cases}$$

2.4 Solutions of Process Models

In previous sections, we have introduced ODE and PDE models and their classification for industrial processes. We have also discussed several examples of ODE and PDE models. Now, we are ready to discuss how to solve ODE and PDE process models.

In the following, three subsections are dedicated to solutions of the initial value problem, the boundary value problem and PDEs, respectively. For some simple ODEs and PDEs, it is possible to derive analytical solutions. However, deriving analytical solutions is difficult for general ODE and PDE models, demanding numerical computation for solutions.

2.4.1 *Solution of an Initial Value Problem of ODE*

We use an example to illustrate what a solution of an initial value problem is. Consider a special case of Eq. (2.6) with $n = 1$

$$a_1 \frac{dy(t)}{dt} + a_0 y(t) = f(t, y(t)), \tag{2.15}$$

with the initial condition

$$y(t = t_0) = y_0 \tag{2.16}$$

A solution to this initial value problem is a function $y(t)$, which satisfies Eqs. (2.15) and (2.16). Similarly, this definition can be easily extended to the general case of (2.6) or a system of ODEs (2.11).

For instance, a solution to the enzyme reaction model in Equation (2.11) is a vector function $(s(t), c(t))$, which satisfies Equation (2.11) and the initial conditions $s(0) = s_0$ and $c(0) = 0$.

2.4.2 *Solution of a Boundary Value Problem of ODE*

Let us consider a special case of Equation (2.6). Letting $a_2 = a_0 = 1, a_1 = 0$ and $f(x, y) = 0$ yields a model that describes the Strum-Liouville problem

$$\frac{d^2 y(t)}{dx^2} + y(t) = 0, x \in [0, \pi/2] \tag{2.17}$$

with the boundary conditions

$$y(0) = 0 \text{ and } y(\pi/2) = 2 \tag{2.18}$$

A function $y(t)$ is called a solution to the boundary value problem if it satisfies Equations (2.17) and (2.18) simultaneously. For example,

$$y(t) = 2\sin(t) \tag{2.19}$$

is an analytical solution to the problem of interest.

2.4.3 *Solution of a PDE*

Consider population balance Equation (2.13). For a crystallization process in a batch vessel with the independent growth, the population balance equation gives a special form as follows

$$\frac{\partial n(t, x)}{\partial t} + G \frac{\partial n(t, x)}{\partial x} = 0 \tag{2.20}$$

and initial and boundary conditions

$$n(0,x) = \exp\left(-\frac{100}{6.6}(x-1)^2\right), \; n(t,0) = \exp\left(-\frac{100}{6.6}(t+1)^2\right). \quad (2.21)$$

Then, a solution to Equation (2.20) with initial and boundary conditions in Equation (2.21) is a function $n(t,x)$ that satisfies Equations (2.20) and (2.21). For example,

$$n(t,x) = \exp\left(-\frac{100}{6.6}(x-t-1)^2\right) \quad (2.22)$$

is an analytical solution to Equations (2.20) and (2.21) for the case of $G = 1$.

It is seen from the last section that analytical solutions can be found for simple ODEs and PDEs. Equation (2.19) is an analytical solutions to Equation (2.17) under the initial condition Equation (2.18); while PDE (2.20) has an analytical solution (2.22) under the boundary conditions shown in Equation (2.21).

While it might be possible to derive analytical solutions for simple ODEs and PDEs, it is difficult to find solutions to general ODE and PDE models. This requires numerical computation of the mathematical models. In the next few sections, we will introduce several commonly used numerical techniques for ODE and PDE models in process systems engineering. The Runge-Kutta methods and finite difference methods are discussed in Sections 2.5 and 2.6. Later in Sections 2.7 and 2.8, we will discuss wavelet-based methods and high resolution technique methods for numerical computation of mathematical models for complex industrial processes.

2.5 The Runge-Kutta Methods

In numerical computation, the Runge-Kutta methods are an important family of implicit and explicit iterative methods, which are used in temporal discretization for the approximation of solutions of ODEs. Many well known numerical schemes, such as the Euler method, the Backward Euler method, and the Trapezoidal method belong to this family of numerical methods. The Runge-Kutta method of order 4 is one of the most popular techniques in this family due to its accuracy, stability and ease to program. Therefore, we will discuss the fourth-order Runge-Kutta method in more detail in the rest of this section.

Consider a general ODE

$$\frac{dy(t)}{dt} = f(t, y(t)), \tag{2.23}$$

Let $y(t)$ be a solution we are seeking to ODE (2.23), the general formula of the Runge-Kutta method can be written as

$$y_{i+1} = y_i + \sum_{j=1}^{\kappa} w_j K_j, \tag{2.24}$$

where κ is the order of the method, y_i is the value of y at ith step, and K_j and the coefficients w_j are determined by the information from previous steps.

For the fourth-order Runge-Kutta method, we have $\kappa = 4$,

$$w_1 = w_4 = \frac{1}{6}, w_2 = w_3 = \frac{1}{3} \tag{2.25}$$

and

$$\begin{cases} K_1 = hf(t_i, y_i), \\ K_2 = hf(t_i + a_1 h, y_i + b_1 K_1), \\ K_3 = hf(t_i + a_2 h, y_i + b_2 K_1 + b_3 K_2), \\ K_4 = hf(t_i + a_3 h, y_i + b_4 K_1 + b_5 K_2 + b_6 K_3) \end{cases} \tag{2.26}$$

where $h > 0$ is a step size, a_i and b_i are given by

$$\begin{aligned} &a_1 = \frac{1}{2}, a_2 = \frac{1}{2}, a_3 = 1, \\ &b_1 = \frac{1}{2}, b_2 = 0, b_3 = \frac{1}{2}, b_4 = 0, b_5 = 0, b_6 = 1. \end{aligned} \tag{2.27}$$

2.6 Finite Difference Methods

In various schemes developed for numerically solving differential equations, the way of formulating the approximation of derivatives is crucial. Finite difference methods are a class of numerical computation methods for approximating the solutions to differential equations using finite difference equations to approximate derivatives.

Assume that the function $f(x)$ whose derivatives are to be approximated is properly-behaved. With a small step size $h > 0$, Taylor series expansions of $f(x \pm h)$ at x have the following forms

$$f(x+h) = f(x) + \frac{df(x)}{dx}h + \frac{df^{(2)}(x)}{dx^2}\frac{h^2}{2} + \frac{df^{(3)}(x)}{dx^3}\frac{h^3}{3!} + \cdots \tag{2.28a}$$

$$f(x-h) = f(x) - \frac{df(x)}{dx}h + \frac{df^{(2)}(x)}{dx^2}\frac{h^2}{2} - \frac{df^{(3)}(x)}{dx^3}\frac{h^3}{3!} + \cdots \tag{2.28b}$$

We have

$$\frac{df(x)}{dx} = \frac{f(x+h) - f(x)}{h} + O(h) \tag{2.29a}$$

$$\frac{df(x)}{dx} = \frac{f(x) - f(x-h)}{h} + O(h) \tag{2.29b}$$

which give two different ways of approximating the first-order derivative of function $f(x)$

$$\frac{df(x)}{dx} = \frac{f(x+h) - f(x)}{h} \tag{2.30a}$$

$$\frac{df(x)}{dx} = \frac{f(x) - f(x-h)}{h} \tag{2.30b}$$

with a first-order truncation error $O(h)$. The derivative approximation in Equation (2.30a) is called the **Forward Difference**, while the derivative approximation in Equation (2.30b) is called the **Backward Difference**.

Combining Equations (2.29a) and (2.29b) leads the **Central Difference approximation** with a second-order truncation error $O(h^2)$ for the first-order derivative of $f(x)$:

$$\frac{df(x)}{dx} = \frac{f(x+h) - f(x-h)}{2h}. \tag{2.31}$$

The finite difference approximation for the second-order derivative of $f(x)$ with second-order truncation error is given by

$$\frac{df^{(2)}(x)}{dx^2} = \frac{f(x+h) - 2f(x) + f(x-h)}{2h} \tag{2.32}$$

Similar idea can be used for approximations of higher-order derivatives of $f(x)$.

With the finite difference approximations developed in Equations (2.30) and (2.31), we explain below how to apply these approximations to ODEs. The idea can be easily extend to the other cases of a set of ODEs.

Consider a process model describing a chemical reaction in a fixed bed. The model equation is given by Equation (2.6) with $x \in [a, b]$ and

$$n = 2, n_2 = 1, a_1 = -2, a_0 = -10, f(x, y) = 0 \tag{2.33}$$

Divide interval $[a, b]$ by x_i into N subintervals with $a = x_0 < x_1 < x_2 < \cdots < x_{N-1} < x_N$ and $\Delta x_i = x_{i+1} - x_i$. Denote $y(x_i)$ by y_i. For example, if a uniform step size is employed and both the first- and second-order derivatives are approximated by forward finite difference scheme, then for ODE (2.6) we have the following approximation

$$\frac{y_{i+2} - 2y_{i+1} + y_i}{\Delta x^2} - 2\frac{y_{i+1} - y_i}{\Delta x} - 10y_i = 0. \tag{2.34}$$

Rearranging Equation (2.34) gives the following finite difference equation

$$y_{i+2} - \alpha y_{i+1} + \beta u_i = 0, \tag{2.35}$$

$$\alpha = 2(1 + \Delta x),\ \beta = 1 + 2\Delta x - 10\Delta x^2. \tag{2.36}$$

This finite difference equation can be solved when boundary conditions are specified.

From the Taylor series expansion, higher order derivatives can also be obtained with various order of accuracy. In order to easily find the difference formulas for different order of derivatives with different order of accuracy, the difference formulas for the first $4th$ order derivatives with up to $4th$ order accuracy are summarized in the Tables 2.2 to 2.5. For simplicity, a uniform step size, h, is used. Then, the step size Δx_i is replaced by h. Note that the numbers in Tables 2.2 to 2.5 represent the corresponding coefficients of u_i.

Table 2.2 Finite difference approximations for $h\dfrac{\partial u}{\partial x}$ at x_i.

	Forward Finite Difference		Backward Finite Difference		Central Finite Difference	
u_{i-2}				1/2		1/12
u_{i-1}			-1	-2	-1/2	-2/3
u_i	-1	-3/2	1	3/2		
u_{i+1}	1	2			1/2	2/3
u_{i+2}		-1/2				-1/12
Error	$O(h)$	$O(h^2)$	$O(h)$	$O(h^2)$	$O(h^2)$	$O(h^4)$

Table 2.3 Finite difference approximations for $h^2\dfrac{\partial^2 u}{\partial x^2}$ at x_i.

	Forward Finite Difference		Backward Finite Difference		Central Finite Difference	
u_{i-3}			1	-1		
u_{i-2}			1	4		-1/12
u_{i-1}			-2	-5	1	16/12
u_i	1	2	1	2	-2	-30/12
u_{i+1}	-2	-5			1	16/12
u_{i+2}	1	4				-1/12
u_{i+3}		-1				
Error	$O(h)$	$O(h^2)$	$O(h)$	$O(h^2)$	$O(h^2)$	$O(h^4)$

Table 2.4 Finite difference approximations for $h^3 \dfrac{\partial^3 u}{\partial x^3}$ at x_i.

	Forward Finite Difference		Backward Finite Difference		Central Finite Difference	
u_{i-4}				3/2		
u_{i-3}			-1	-7		1/8
u_{i-2}			3	12	-1/2	-1
u_{i-1}			-3	-9	1	13/8
u_i	-1	-5/2	1	5/2		
u_{i+1}	3	9			1	-13/8
u_{i+2}	-3	-12			1/2	1
u_{i+3}	1	7				-1/8
u_{i+4}		-3/2				
Error	$O(h)$	$O(h^2)$	$O(h)$	$O(h^2)$	$O(h^2)$	$O(h^4)$

Table 2.5 Finite difference approximations for $h^4 \dfrac{\partial^4 u}{\partial x^4}$ at x_i.

	Forward Finite Difference		Backward Finite Difference		Central Finite Difference	
u_{i-5}				-2		
u_{i-4}			-1	11		
u_{i-3}			-4	-24		-1/6
u_{i-2}			6	26	1	2
u_{i-1}			-4	-14	-4	39/6
u_i	1	3	1	3	6	56/6
u_{i+1}	-4	-14			-4	-39/6
u_{i+2}	6	26			1	2
u_{i+3}	-4	-24				-1/6
u_{i+4}	1	11				
u_{i+5}		-2				
Error	$O(h)$	$O(h^2)$	$O(h)$	$O(h^2)$	$O(h^2)$	$O(h^4)$

For example, we use forward finite difference method to approximate $\dfrac{\partial u}{\partial x}$. From Table (2.2), we have

$$\frac{\partial u}{\partial x}\Big|_{x=x_i} = \frac{(1) \times u_{i+1} + (-1) \times u_i}{h} \tag{2.37}$$

for an accuracy of first order. For an accuracy of second order, we have

$$\frac{\partial u}{\partial x}\Big|_{x=x_i} = \frac{(-\frac{3}{2}) \times u_i + (2) \times u_{i+1} + (-\frac{1}{2}) \times u_{i+2}}{h}. \tag{2.38}$$

If a process model is described by a PDE, a multidimensional finite difference scheme can be employed for numerical approximation. This can be easily derived from the one-dimensional forward finite difference scheme developed previously by using Tables 2.2 to 2.5. For example, applying the central finite difference scheme to population balance Equation (2.20) gives

$$\frac{n_{i+1,j} - n_{i-1,j}}{2\Delta t_i} + G\frac{n_{i,j+1} - n_{i,j-1}}{2\Delta x_i} = 0 \tag{2.39}$$

where the indices i and j are for t-direction and x-direction, respectively.

2.7 Wavelets-Based Methods

Functions with fast oscillations, or even discontinuities, in localized regions are commonly met in engineering. How to capture the essential behavior of the processes with such dynamics is crucial in developing an algorithm for numerical computation of the process models. Traditional methods, such as the Fourier expansion, must use many basis functions to approximate those process functions due to the infinite support in the mathematical models.

Wavelet analysis is a relatively new mathematical technique for complex mathematical models. It has attracted much interest in both theoretical and applied mathematics over the past decade, particularly in numerical analysis of mathematical models.

In contrast with traditional techniques for numerical computation, wavelets may have compact support, enabling approximation of a function not by cancellation, but through placement of the right wavelets at an appropriate location. They are cable of analyzing different parts of a function at different scales, and can also represent polynomials up to a certain order exactly, leading to some very successful applications in numerical analysis of mathematical models.

In the rest of this section, we will introduce multiresolution analysis first. Then, we will discuss the popularly used Daubechies' wavelets, including the Haar wavelet, Daubechies orthonormal wavelet and interpolating wavelet. These wavelets will be used in the rest of the book for wavelets-based numerical computation. Furthermore, the computation of the connection coefficients in wavelet-based methods will also be investigated.

2.7.1 *Multiresolution Analysis*

In wavelets-based numerical methods, effective numerical algorithms can be developed from the **MultiResolution Analysis (MRA)** technique. According to [Cohen (2003)], a multiresolution analysis or multiscale approximation of the space $L^2(R)$ is a sequence of closed subspaces of $L^2(R)$ with the following axioms holding:

- The sequence is nested, i.e.,

$$0 \subset \cdots \subset V_{-1} \subset V_0 \subset V_1 \cdots L^2(R) \tag{2.40}$$

- It satisfies the two-scaling relation

$$f \in V_j \Leftrightarrow f(2\cdot) \in V_{j+1}. \tag{2.41}$$

- The union of the spaces is dense, i.e.,

$$\overline{\bigcup_{j=-\infty}^{\infty}} V_j = L^2(R). \tag{2.42}$$

- There is a function ϕ such that

$$\{\phi(\cdot - k)\}_{k \in Z} \tag{2.43}$$

  is a Reisz basis of V_0.

In a slightly different language, a multiresolution analysis of the space $L^2(R)$ consists of a sequence of nested subspaces as shown in Equation (2.40) that satisfies certain self-similarity relations in time/space and scale/frequency, as well as completeness and regularity relations. In mathematics, the completeness demands that those nested subspaces

(1) fill the whole space, i.e., their union should be dense in $L(R)$ as shown in Equation (2.42); and
(2) are not too redundant, i.e., their intersection should only contain the zero element.

If we denote the orthogonal complement of V_j in V_{j+1} by W_j, the basis of W_j can then be obtained by dilating and translating the **scaling function** ϕ as follows

$$\psi = \sum_j c_{j,k} \phi(2 \cdot -k), \tag{2.44}$$

which is called **mother wavelet** or **basic wavelet**. Then, a smooth function, $f(x)$ in $V_J \in L^2(R)$ can be expressed as a scaling function expansion

$$f(x) = \sum_{k=-\infty}^{\infty} f_{J,k} \phi_{J,k}, \tag{2.45}$$

or a wavelet expansion

$$f(x) = \sum_{k=-\infty}^{\infty} f_{J_0,k}\phi_{J_0,k} + \sum_{j=J_0}^{J-1} \sum_{k=-\infty}^{\infty} d_{j,k}\psi_{j,k}, \tag{2.46}$$

where

$$\phi_{j,k} = \phi(2^j \cdot -k), \tag{2.47}$$

$$\psi_{j,k} = \psi(2^j \cdot -k) \tag{2.48}$$

are basis functions of V_j and W_j, respectively; and coefficients $f_{j,k}$ and $d_{j,k}$ are defined by

$$f_{j,k} = \int_{-\infty}^{\infty} f(x)\phi_{j,k}(x)dx, \tag{2.49}$$

$$d_{j,k} = \int_{-\infty}^{\infty} f(x)\psi_{j,k}(x)dx. \tag{2.50}$$

2.7.2 *Basis Functions of Daubechies' Wavelets*

Three types of Daubechies' wavelets and their basis functions will be discussed in this subsection: the Haar wavelet, Daubechies orthonormal wavelet and interpolating wavelet.

The Haar Wavelet

The Haar wavelet is the simplest orthonormal wavelet. It has basis functions with explicit expressions as follows

- The scaling function:

$$\phi(x) = \begin{cases} 1, & 0 \le x < 1 \\ 0, & \text{otherwise} \end{cases} \tag{2.51}$$

- The mother wavelet:

$$\psi(x) = \begin{cases} 1, & 0 \le x < 1/2 \\ -1, & 1/2 \le x < 1 \\ 0, & \text{otherwise} \end{cases} \tag{2.52}$$

Daubechies Orthonormal Wavelet

In 1992, Daubechies constructed a family of compactly supported orthonormal wavelets, which include members from highly localized to highly smooth ones. Although there is no explicit expression for the basis functions of the wavelet, the value of the basis functions at the dyadic point can be obtained from the following two-scale relations both for the scaling function and wavelet:

$$\phi(x) = \sum_{k=0}^{L-1} p_k \phi(2x - k) \tag{2.53}$$

and

$$\psi(x) = \sum_{k=0}^{L-1} (-1)^k p_{1-k} \phi(2x - k) \tag{2.54}$$

with the coefficients $p_k, k = 1, 2, \cdots, L-1$, and the compact support $[0, L-1]$ for ϕ and ψ. The coefficients $p_k, k = 1, 2, \cdots, L-1$, are known as **filter coefficients**. The translations and dilations of level j for $\phi(x)$ and $\psi(x)$ are respectively defined as

$$\phi_{j,k}(x) = 2^{j/2} \phi(2^j x - k) \tag{2.55}$$

and

$$\psi_{j,k}(x) = 2^{j/2} \psi(2^j x - k) \tag{2.56}$$

with compact support $[\frac{k}{2^j}, \frac{k+L-1}{2^j}]$. Then, $\{\phi_{j,k}\}$ and $\{\psi_{j,k}\}$ consist of the basis of V_j and W_j, respectively.

The following three properties can be derived from Equations (2.53) and (2.54):

(1) Polynomials can be represented exactly up to some degree of $P - 1$ by the scaling function:

$$x^p = \sum_{k=-\infty}^{\infty} M_k^p \phi(x - k), \; x \in \mathbb{R}, p = 0, 1, \cdots, P-1 \tag{2.57}$$

with the pth moment of $\phi(x)$ defined by

$$M_k^p = \int_{-\infty}^{\infty} x^p \phi(x - k) dx. \tag{2.58}$$

(2) P vanishing moments for the wavelet:

$$\int_{-\infty}^{\infty} x^p \psi(x) dx = 0, \; x \in \mathbb{R}, p = 0, 1, \cdots, P-1 \tag{2.59}$$

(3) Orthonormality of the filter coefficients:

$$\sum_k p_k p_{k-2n} = 2\delta_{0,n}, \; n \in \mathbb{Z} \tag{2.60}$$

where $\delta_{0,n}$ is the Delta function.

Interpolating Wavelet

In 1992, D. Donoho developed a new type of wavelet, interpolating wavelet, which is developed from existing wavelets. For instance, the one based on Daubechies wavelet is given by

$$\theta(x) = \int_{-\infty}^{\infty} \phi(y)\phi(y-x)dy \tag{2.61}$$

where ϕ is the Duabechies scaling function. θ satisfies the Kronecker property

$$\theta(k) = \begin{cases} 1, & k = 0 \\ 0, & \text{otherwise} \end{cases} \tag{2.62}$$

In addition, θ also satisfies all properties of the general Daubechies wavelets. Thus, functions

$$\{\theta_{j,k}(x)\}_{k=0,1\cdots,2^j} = \{\theta(2^j x - k)\}_{k=0,1\cdots,2^j} \tag{2.63}$$

consist of the basis of V_j; and functions

$$\{\theta_{j+1,2k+1}(x)\}_{k=0,1\cdots,2^j} = \{\theta(2^{j+1}x - 2k - 1)\}_{k=0,1\cdots,2^j} \tag{2.64}$$

consist of the basis of W_j, which is the orthogonal complement of V_j in V_{j+1}.

Either $\phi_{j,k}, \psi_{j,k}$ in Equations (2.55) and (2.56) or $\theta_{j,k}$ in Equations (2.63) and (2.64) can be used as basis functions to approximate a smooth function f as expressed in Equations (2.45) and (2.46). Depending on the way of how to calculate the coefficients $f_{j,k}$ and $d_{j,k}$ and which basis functions are used, wavelet-based numerical techniques can be categorized as the Wavelet Galerkin Method and Wavelet Collocation Method. Both of these two wavelet methods will be discussed later in this book for computation of mathematical models for complex industrial processes.

2.7.3 *Computation of the Connection Coefficients*

Because the basis functions $\theta_{j,k}$ satisfy the Kronecker property, the coefficients appeared in the wavelet collocation method are relatively easy to compute. However, in the wavelet Galerkin method, one of the most difficulties is to calculate the so-called **connection coefficients** defined below in Equation (2.65). Now let us briefly review the algorithm that we recently rectified in [Zhang *et al.* (2007)] to calculate the connection coefficients, namely the integral of the product of the scaling function $\phi(x)$ and its nth derivative $\phi^{(n)}(x-k)$

$$\Gamma_k^n(x) = \int_0^x \phi^{(n)}(y-k)\phi(y)dy. \tag{2.65}$$

Let

$$\Gamma^n(L-1) = \left[\Gamma_0^n(L-1), \Gamma_1^n(L-1), \cdots, \Gamma_{L-2}^n(L-1)\right]^T. \tag{2.66}$$

Then, from [Chen *et al.* (1996)] and [Zhang *et al.* (2007)], we can easily obtain the values of $\Gamma_k^n(L-1)$ through the following relation:

$$\Gamma^n(L-1) = D\Gamma^n(L-1) \tag{2.67}$$

with normalization condition

$$\sum_{k=0}^{L-2} k^n \Gamma_k^n(L-1) = \frac{n!}{2} \tag{2.68}$$

where

$$D = (d_{l,m}),\ l,m = 1,2,\cdots,L-1 \tag{2.69}$$

$$d_{l,m} = 2^{n-1}\left[\sum_{\mu_1(l,\ m)} p_i p_j + (-1)^n \sum_{\mu_2(l,\ m)} p_i p_j\right] \tag{2.70}$$

and

$$\mu_\lambda(l,m) = \{(i,j) : 0 \le i,j \le L-1\ 2(l-1)+i-j = (-1)^{\lambda+1}(m-1)\}, \lambda = 1,2. \tag{2.71}$$

After getting the values of $\Gamma_k^n(x)$ for $x = L-1$, we can compute the values of $\Gamma_k^n(x)$ for $x = 0,1,\cdots,L-2$ and $k = 2-L, 3-L, \cdots, L-2$. In order to do so, let

$$\Gamma^n = [\Gamma^n(1), \cdots, \Gamma^n(L-2)]^T \tag{2.72}$$

with

$$\Gamma^n(i) = [\Gamma_{i-L+2}^n(i), \cdots, \Gamma_{i-1}^n(i)]^T,\ i = 1,2,\cdots,L-2. \tag{2.73}$$

Then, we have the following system for $\Gamma_k^n(x)$ with $x = 1,\cdots,L-2$ and $k = x-L+2,\cdots,x-1$

$$\widetilde{Q}\Gamma^n = (2^{1-n}\widetilde{I} - Q)\Gamma^n = d, \tag{2.74}$$

where $\widetilde{I}$ is a square unit matrix of order $(L-2)^2$, $Q = (Q_{i,j})$ is a square matrix of order $(L-2)^2$ with $Q_{i,j} = (q_{i,j,k,m})$ being a $(L-2)\times(L-2)$ matrix and $q_{i,j,k,m} = p_{2i-j}p_{L-1-2k+m}$, and

$$d = \left[d^1, d^2, \cdots, d^{L-2}\right]^T,$$

with

$$d^i = [d((i-1)(L-2)+1), \cdots, d((i-1)(L-2)+k), \cdots, d(i(L-2)]^T, \tag{2.75a}$$

$$d((i-1)(L-2)+k) = \sum_{\mu_2(i,k,L)} p_{i_1} p_{j_1} \Gamma^n_{2(i-(L-2)+(k-1))+i_1-j_1}(L-1), \tag{2.75b}$$

$$\mu_2(i,k,L) = \{(i_1, j_1) \in \mu(i,k,L) : 2i - j_1 \geq L-1 \text{ or } 2\mathrm{k} + \mathrm{i}_1 \leq \mathrm{L} - 1\}. \tag{2.75c}$$

The value of Γ^n for $n = 0$ can then be easily obtained from Equation (2.74). For the case of $n > 0$, the following vector equation is required

$$[(x-L+2)^n, \cdots, (x-1)^n]\, \Gamma^n(x) = n!\theta_1(x) - \sum_{l=L-1-x}^{L-2} l^n \Gamma^n_l(L-1). \tag{2.76}$$

Then, for $i = 1, 2, \cdots, n$, doing the following two steps gives the value of Γ^n for $n > 0$:

(1) Replace the ith row of $\widetilde{Q}_{i,i} = 2^{1-n}I - Q_{i,i}$ and $\widetilde{Q}_{i,j} = -Q_{i,j}$ by $[(i-L+2)^n, \cdots, (i-1)^n]$ and a zero row vector of order $L-2$, respectively;
(2) Replace $d((i-1)(L-2)+i)$, the ith element of d^i, by $n!\theta_1(i) - \sum_{l=L-1-i}^{L-2} l^n \Gamma^n_l(L-1)$.

Now, we have calculated all connection coefficients as defined in Equation (2.65).

2.8 High Resolution Methods

In 1993, Koren posed an algorithm for numerically solving PDEs with Dirichlet boundary condition. It is now known as **High Resolution (HR) method**. Later on, this method has been adopted for solving population balance equations (PBEs) with Dirichlet boundary conditions [Gunawan *et al.* (2004); Qamar *et al.* (2006)]. It has also been developed for solving PDEs with Cauchy or Neumann boundary conditions [Zhang *et al.* (2008)].

This section briefly introduces this numerical computation method. The method will be further applied later in this book for computation of mathematical models of complex industrial processes.

2.8.1 *Koren's High-Resolution Scheme*

Consider the following PDE

$$\frac{\partial u}{\partial t} + \frac{\partial f(u)}{\partial x} + \beta \frac{\partial^2 u}{\partial x^2} = 0. \tag{2.77}$$

Divide the space interval $[a, b]$ for x into N subintervals $\Omega_i = [x_{i-1/2}, x_{i+1/2}], i = 1, \cdots, N$, with $x_{1/2} = a$ and $x_{N+1/2} = b$. Let $x_i = \dfrac{x_{i-1/2} + x_{i+1/2}}{2}$ and $\Delta x_i = x_{i+1/2} - x_{i-1/2}$. It follows that $x_i = x_{i-1/2} + \Delta x_i/2$. Then, the unknown u in Ω_i can be approximated as follows

$$u_i(t) = \frac{1}{\Delta x_i} \int_{x_{i-1/2}}^{x_{i+1/2}} u(x, t) dx. \tag{2.78}$$

Integrating Equation (2.77) on both sides for x from $x_{i-1/2}$ to $x_{i+1/2}$ gives the following semi-discrete equation

$$\frac{\partial u_i(t)}{\partial t} + \frac{1}{\Delta x_i}(f_{i+1/2} - f_{i-1/2}) + \frac{1}{\Delta x_i}\beta \left(\frac{\partial u}{\partial x}\Big|_{x_{i+1/2}} - \frac{\partial u}{\partial x}\Big|_{x_{i-1/2}} \right) = 0. \tag{2.79}$$

In the following, we will discuss how Equation (2.79) can be solved numerically.

Approximation of $f_{i\pm1/2}$

There are two basic ways to approximate $f_{i\pm1/2}$: upwind scheme and $\kappa-$flux interpolation scheme. The upwind scheme is a first-order approximation, which is given by

$$f_{i+1/2} = f_i, \tag{2.80}$$

while the $\kappa-$flux interpolation scheme is represented by

$$f_{i+1/2} = f_i + \frac{1+\kappa}{4}(f_{i+1} - f_i) + \frac{1-\kappa}{4}(f_i - f_{i-1}),\ \kappa \in [-1, 1]. \tag{2.81}$$

In the $\kappa-$flux interpolation scheme, we are especially interested in the case of $\kappa = 1/3$, with which the optimized $\kappa-$interpolation approximation may improve the accuracy of numerical computation. For $\kappa = 1/3$, we have

$$f_{i+1/2} = f_i + \frac{1}{2}\left(\frac{1}{3} + \frac{2}{3} r_i^+\right)(f_i - f_{i-1}), \tag{2.82}$$

or

$$f_{i+1/2} = f_i + \frac{1}{2}\Phi(r_i^+)(f_i - f_{i-1}), \tag{2.83}$$

where the flux limited function Φ is defined by

$$\Phi(r) = \max\left(0, \min\left(2r, \min\left(\frac{1}{3} + \frac{2}{3}r, 2\right)\right)\right). \tag{2.84}$$

The upwind ratio of two consecutive flux gradients is defined by

$$r_i^+ = \frac{f_{i+1} - f_i + \epsilon}{f_i - f_{i-1} + \epsilon}, \tag{2.85}$$

where $\epsilon > 0$ is a small parameter to avoid division by zero.

Approximation of $\frac{\partial u}{\partial x}\Big|_{x_{i+1/2}}$

$\frac{\partial u}{\partial x}\Big|_{x_{i+1/2}}$ can be approximated by either forward or backward difference approximation. For forward difference approximation, we have

$$\frac{\partial u}{\partial x}\Big|_{x_{i+1/2}} = \frac{u_{i+1} - u_i}{\Delta x_i}, \tag{2.86}$$

while for backward difference approximation, we have

$$\frac{\partial u}{\partial x}\Big|_{x_{i-1/2}} = \frac{u_i - u_{i-1}}{\Delta x_i}. \tag{2.87}$$

It is not surprising that all those approximations work well at internal points, i.e., for $\Omega_i, i = 2, \cdots, N-1$.

In the following subsections, we will deal with different boundary conditions for the PDE model in Equation (2.77) or its semi-discrete form in Equation (2.79).

2.8.2 *Solving PDEs with Dirichlet Boundary Conditions*

Assume the Dirichlet boundary conditions

$$u(t, a) = u_{in}(t) =: u_{in}, \tag{2.88a}$$

$$u(t, b) = u_{out}(t) =: u_{out}, \tag{2.88b}$$

are held for PDE (2.77). Then, we have exact values for $f_{1/2}$ and $f_{N+1/2}$ at the boundaries

$$f_{1/2} = f(u(t, a)) = f(u_{in}), \tag{2.89a}$$

$$f_{N+1/2} = f(u(t, b)) = f(u_{out}). \tag{2.89b}$$

For $i = 1$, Equation (2.81) is only valid for $\kappa = 1$. Thus, the 1−flux interpolation scheme gives $f_{3/2}$ as follows

$$f_{3/2} = \frac{f_1 + f_2}{2} \text{ or f} \left(\frac{\mathrm{u}_1 + \mathrm{u}_2}{2}\right). \tag{2.90}$$

As there is no information on $\frac{\partial u}{\partial x}$ from the boundary conditions in this case, the biased second-order accuracy difference will be employed to approximate the values of $\frac{\partial u}{\partial x}|_{i+1/2}$ at the left and right ends of the interval $[a, b]$ for x. This gives

$$\frac{\partial u}{\partial x}|_{1/2} = \frac{-8u(t,a) + 9u_1 - u_2}{3\Delta x_0}, \tag{2.91}$$

and

$$\frac{\partial u}{\partial x}|_{N+1/2} = \frac{8u(t,b) - 9u_N + u_{N-1}}{3\Delta x_N}. \tag{2.92}$$

2.8.3 *Solving PDEs with Cauchy Boundary Conditions*

In this section, Koren's scheme is extended for numerically solving PDEs with Cauchy boundary conditions.

Assuming the PDE model in Equation (2.77) has the following Cauchy boundary conditions:

$$u(t,a) + \alpha \frac{\partial u(t,x)}{\partial x}|_{x=a} = u_{in}(t) =: u_{in}, \tag{2.93a}$$

$$\frac{\partial u(t,x)}{\partial x}|_{x=b} = u_{out}(t) =: u_{out}. \tag{2.93b}$$

In the internal subintervals, $\Omega_i, i = 2, \cdots, N-1$, the formulae given in previous section 2.8.1 can still be used to approximate the unknowns. In the remaining part of this section, we develop approximations on the boundaries.

Recall that the biased second-order accuracy difference at $x = a$ is given by

$$\frac{\partial u}{\partial x}|_{x=a} = \frac{-8u(t,a) + 9u_1 - u_2}{3\Delta x_1}. \tag{2.94}$$

Then, we have

$$u(t,a) = \frac{3\Delta x_1 u_{in} - 9\alpha u_1 + \alpha u_2}{3\Delta x_1 - 8\alpha}, \tag{2.95}$$

which gives the following approximations for f and $\frac{\partial u}{\partial x}$ at $x = a$:

$$f|_{1/2} = f(u(t,a) + \frac{1}{2}(u_1 - u(t,a))), \text{ or} \tag{2.96a}$$

$$f|_{1/2} = f(u(t,a)) + \frac{1}{2}(f(u_1) - f(u(t,a))), \tag{2.96b}$$

and

$$\frac{\partial u}{\partial x}\mid_{1/2}=\frac{-8u_{in}+9u_1-u_2}{3\Delta x_1-8\alpha}. \tag{2.97}$$

Because of the same reason as mentioned in the last section, for $i=1$, we have

$$f_{3/2}=\frac{f_1+f_2}{2} \text{ or } \mathrm{f}\left(\frac{\mathrm{u}_1+\mathrm{u}_2}{2}\right). \tag{2.98}$$

Applying the -1-interpolation formula instead of the general $\kappa-$interpolation for f and $\frac{\partial u}{\partial x}$ at $x=b$ yields

$$f\mid_{N+1/2}=f(u_N+\frac{1}{2}(u_N-u_{N-1})), \text{ or} \tag{2.99a}$$

$$f\mid_{N+1/2}=f(u_N)+\frac{1}{2}(f(u_N)-f(u_{N-1})), \tag{2.99b}$$

and

$$\frac{\partial u}{\partial x}\mid_{N+1/2}=u_{out}. \tag{2.100}$$

2.8.4 *Solving PDEs with Neumann Boundary Conditions*

This section extends Koren's scheme for numerically solving PDEs with Neumann boundary conditions.

Assume that the Neumann boundary conditions hold for PDE (2.77):

$$\frac{\partial u(t,x)}{\partial x}\mid_{x=a}=u_{in}(t)=: u_{in}, \tag{2.101a}$$

$$\frac{\partial u(t,x)}{\partial x}\mid_{x=b}=u_{out}(t)=: u_{out}. \tag{2.101b}$$

Similar to the derivation in Section 2.8.3, the formulae given in Section 2.8.1 are used to approximate the unknowns in the internal subintervals, $\Omega_i, i=2,\cdots,N-1$. Using the biased second-order accuracy difference at $x=a$ again gives

$$u(t,a)=\frac{-3\Delta x_1 u_{in}+9u_1-u_2}{8}, \tag{2.102}$$

which gives the following approximations for f at $x=a$

$$f\mid_{1/2}=f(u(t,a)+\frac{1}{2}(u_1-u(t,a))), \text{ or} \tag{2.103a}$$

$$f\mid_{1/2}=f(u(t,a))+\frac{1}{2}(f(u_1)-f(u(t,a))). \tag{2.103b}$$

For $i = 1, N$, we have

$$f_{3/2} = \frac{f_1 + f_2}{2} \text{ or f} \left(\frac{\mathrm{u}_1 + \mathrm{u}_2}{2} \right), \tag{2.104}$$

$$f\ |_{N+1/2} = f(u_N + \frac{1}{2}(u_N - u_{N-1})), \text{ or} \tag{2.105a}$$

$$f\ |_{N+1/2} = f(u_N) + \frac{1}{2}(f(u_N) - f(u_{N-1})), \tag{2.105b}$$

and

$$\frac{\partial u}{\partial x}\ |_{N+1/2} = u_{out}. \tag{2.106}$$

Chapter 3

Finite Difference Methods for Ordinary Differential Equation Models

This chapter is dedicated to development of finite difference methods for numerical computation of ordinary differential equation (ODE) models established for complex industrial processes. Biological fermentation processes, which are industrially significant, are taken as an example for process modelling, resulting in a typical initial value problem in ODEs. Then, the initial value problem in ODEs is solved numerically using the finite difference technique. Simulation studies are carried out to demonstrate numerical computation of the ODE model for a fermentation process.

The content presented in this chapter is partially taken from our previous publications [Yao *et al.* (2001); Tian *et al.* (2002)].

3.1 Fermentation Processes

Fermentation processes have been used since early civilizations, and have been continually improved with time. Nowadays, they are widely used in chemical, pharmaceutical, and other industries.

Chemicals produced by fermentation number about 200. These chemicals can be categorized into two general types: commodity chemicals and specialty chemicals. The former includes ethanol, MSG and citric acid; while the latter includes itaconic acid, antibiotics and enzymes.

Fermentation is often considered as a type of chemical manufacturing. It is also equally well characterized as microbial farming. Historically, fermentation technologies are evolved to the production of increasing valuable products such as antibiotics. More recently, the advent of recombinant DNA technology has dictated development of novel processes for production of totally synthetic products. Scaling up of these novel fermentation processes is also significant in industrial applications. The incorporation of modern engineering, concepts and methods in development, design and control of fermentation operation is becoming universal.

Most fermentation processes are carried out as batch or fed-batch operations. In batch operation, all ingredients are fed to a fermentation vessel at the beginning of the operation. Then, no addition or withdraw of materials takes place during the run of the fermentation operation. In comparison, fed-batch operation involves addition of materials, usually carbon source.

In this chapter, we will use lysine fermentation as a case study. We will demonstrate how fermentation processes can be modeled. The process modelling results in a typical initial value problem in ODEs. Then, we will show how the initial value problem in ODEs for the fermentation process can be numerically solved using the finite difference technique.

It is worth mentioning that for batch fermentation, the model structure and kinetic parameters are adopted from literature [Klasson *et al.* (1991)], where the representation of model is relatively simple and an analytical solution exists.

For fed-batch fermentation, a more comprehensive model is chosen from one of our previous projects [Yao *et al.* (2001)], where concentration variables are interacted. Then, the fed-batch fermentation process under consideration is simulated under two feeding patterns: the pulse feeding and the step feeding. Both cases have employed a mutant strain of Brevibacterium lactofermentum to produce lysine. The fermentation mechanisms in both cases are similar.

To better understand the modelling and model computation for the fermentation process, let us briefly introduce the biology of lysine synthesis in the next section.

3.2 Biology of Lysine Synthesis

Dietary sources of some amino acids are required by human and animals for healthy growth and development. A lack of one or more essential amino acids in diet may give rise to characteristic deficiency diseases.

Lysine is one of the twenty-two common amino acids found in protein. It is not biologically synthesized by the body itself, or at least not sufficient to meet the body's needs for normal growth or maintenance. Its content in grains except soybean, oilseed meal and other feed materials is also very low. Therefore, it becomes vital to add lysine for food enrichment to improve healthy growth and tissue synthesis for human and animals.

Producing lysine through fermentation can be tracked back to 1957. It has been recognized that there exists in all microbial cells a pool or reservoir of amino acids. Although amino acids are essentially transient intermediates in protein synthesis, free amino acids may be excreted into medium, particularly towards the end of the exponential phase of growth when protein synthesis is on the decline.

Lysine is produced by fermentation using cane molasses, glucose, acetic acid or ethanol as the carbon source and ammonium salts or urea as the nitrogen source. In a suitable nutrient medium, micro-organisms extract nutrients from the medium and convert them into biochemical compounds. Part of these nutrients is used for energy production and part is used for biosynthesis and product formation. The whole process involves the following stages:

Stage 1: Cell growth exponentially until the supplied threonine is exhausted, at which time Lysine production commences due to the loss of concerted feedback inhibition of aspartakinase.

Stage 2: Lysine production commences and cell growth continues.

Stage 3: Growth enters a stationary period, during which lysine synthesis continues.

Stage 4: The culture progresses into a slow death phase, where lysine synthesis diminishes with time.

Figure 3.1 shows the above process of cell growth, the substrate depiction and production formation. It can be seen that lysine fermentation belongs to a mixed-growth-associated product whose formation takes place during the slow growth and stationary phases.

3.3 Model Construction

In general, homogeneity is assumed within the fermenter for the sake of simplicity. A suitable modeling approach is to develop a 'balanced' mathematical model of the system under study. This technique is based on the combined use of an assumed kinetic model and a material balance for a single chemical constituent.

The common state variables under consideration are biomass (X),

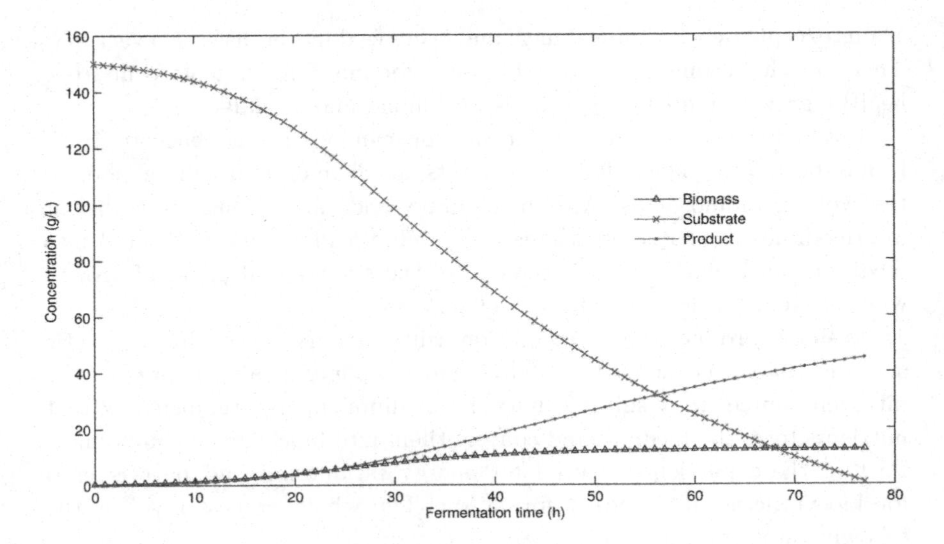

Fig. 3.1 Time profile of lysine batch fermentation.

substrate (S), and product (P) concentration in the liquid medium. The model format used was proposed in [Pirt (1975)] as an ODE:

$$-\frac{dS}{dt} = m_x X + \frac{1}{Y_{x/s}}\frac{dX}{dt} + \frac{1}{Y_{p/s}}\frac{dP}{dt}. \tag{3.1}$$

In model Equation (3.1), the left-hand side represents the consumption rate of a given substrate; while the right hand side is composed of three terms. On the right-hand side, the first term represents non-growth associated substrate use, i.e., for maintenance of cell viability. The second term stands for the substrate consumption required for cell growth. The third term is the substrate used for product formation.

When the material balance model is used with product and biomass, and considering substrate utilized for product formation, the following relationships can be established:

Batch Model:

$$dX/dt = r_x, \tag{3.2}$$

$$dP/dt = r_p, \tag{3.3}$$

$$dS/dt = -r_s, \tag{3.4}$$

where r_x, r_p and r_s are rates of cell growth, carbon substrate consumption and product formation, respectively.

Fed-Batch Model:

$$dX/dt = r_x - X(F/V), \tag{3.5}$$
$$dP/dt = r_p - P(P/V), \tag{3.6}$$
$$dS/dt = -r_s + (S_f - S)(F/V), \tag{3.7}$$
$$dV/dt = F, \tag{3.8}$$

where V is the volume of the fermenter, F stands for the flow rate of the feeding in volume, and S_f represents the substrate concentration in the feeding.

The rate variables r_x, r_p and r_s are characterized by rate equations expressed by kinetic parameters of the fermentation process. For example, the following formulae are commonly adopted as rate equations:

$$r_x = \mu X, \tag{3.9}$$
$$r_p = \alpha r_x + \beta X, \tag{3.10}$$
$$r_s = \frac{r_x}{Y_{x/s}} + \frac{r_p}{Y_{p/s}} + m_s X, \tag{3.11}$$

where μ stands for the specific growth rate of variable cells; Y_p/s and Y_x/s are yield coefficients for product and cells on carbon substrate, respectively; m_s represents maintenance coefficient for cells on carbon substrate.

It is worth mentioning that for batch fermentation, a constant fermenter volume is maintained and the addition of alkali for pH adjustment can be neglected for dynamic modelling. This implies that F/V=0.

3.4 Numerical Approximations of Fermentation Models

This section considers numerical approximations of batch fermentation model equations developed in Section 3.3. With these approximations, numerical computation of the process models becomes possible.

Rate Equation (3.9) is usually approximated by using the relationships derived from the theory of enzymic or chemical reaction. A simple relationship is chosen here so that analytical solutions can be derived for evaluation of proposed numerical solver performance. More complicated rate equations can also be considered for better model accuracy although obtaining an analytical solution is impossible in many cases.

Let us start with rate Equations (3.9), (3.10) and (3.11) of the fermentation process model. Using the growth model from reference

[Klasson *et al.* (1991)], the specific growth rate μ in Equation (3.9) is expressed as:

$$\mu = \mu_m \left(1 - \frac{X}{X_{\max}}\right). \tag{3.12}$$

This is a Riccati equation, which predicts that the specific growth rate is zero once the maximum biomass concentration is reached. It also describes the slowdown behaviour of the growth rate when the cell concentration approaches the maximum. However, it does not relate the growth to the carbon source.

For rate Equation (3.11), an re-arrangement of the equation gives

$$r_s = Cr_x + DX \tag{3.13}$$

where

$$C = \frac{1}{Y_{x/s}} + \frac{\alpha}{Y_{p/s}}, D = \frac{\beta}{Y_{p/s}} + m_s. \tag{3.14}$$

Substituting Equations (3.10), (3.12) and (3.13) into material balance Equations (3.2) - (3.4) yielding a batch fermentation process model

$$\frac{dX(t)}{dt} = \mu_m \left(1 - \frac{X}{X_{\text{Max}}}\right) X(t), \tag{3.15}$$

$$\frac{dP(t)}{dt} = \alpha \frac{dX(t)}{dt} + \beta X(t), \tag{3.16}$$

$$\frac{dS(t)}{dt} = -C\frac{dX(t)}{dt} - DX(t). \tag{3.17}$$

This model is an initial value problem represented by a set of ordinary differential equations with some initial conditions, e.g., $X(0) = X_0, P(0) = P_0$ and $S(0) = S_0$, from the fermentation process operation.

The initial value problem of ODEs (3.15) - (3.17) has analytical solutions as follows:

$$X(t) = \frac{X_{\text{Max}} X_0 e^{\mu_m t}}{X_{\text{Max}} - X_0 + X_0 e^{\mu_m t}}, \tag{3.18}$$

$$P(t) = \alpha(X - X_0) + \frac{\beta X_{\text{Max}}}{\mu_m \ln\left(\dfrac{X_{\text{Max}} - X_0 + X_0 e^{\mu_m t}}{X_{\text{Max}}}\right)}, \tag{3.19}$$

$$S(t) = S_0 - C(X - X_0) - \frac{DX_{\text{Max}}}{\mu_m \ln\left(\dfrac{X_{\text{Max}} - X_0 + X_0 e^{\mu_m t}}{X_{\text{Max}}}\right)}. \tag{3.20}$$

For numerically solving differential equations, there are generally two basic approaches. The one is to use independent functions and the other is to employ finite difference methods. When finite difference methods are used, the solution is approximated by its value at a sequence of discrete points. This is a step by step method, where a rule is provided for computing the approximation at current step in terms of the values obtained at preceding points.

In the time axis, we have a sequence of N discrete points or mesh points $t_0, t_1, \cdots, t_{N-1}$. These points are equally spaced with the step size of $h = t_{i+1} - t_i$. Let $\mathbf{y}_i$ be an approximation of the state variable vector $[X, P, S]^T$ at t_i, and $\mathbf{f}$ be a vector of the derivatives of X, P and S, respectively, i.e.,

$$\mathbf{y}_i = [X(t_i), P(t_i), S(t_i)]^T, \tag{3.21}$$

$$\mathbf{f} = \left[\frac{dX}{dt}, \frac{dP}{dt}, \frac{dS}{dt}\right]^T. \tag{3.22}$$

Then, the general formula of the Runge-Kutta methods can be written as:

$$\mathbf{y}_{i+1} = \mathbf{y}_i + \sum_{j=1}^{\kappa} w_j K_j \tag{3.23}$$

where

$$K_j = h\mathbf{f}\left(t_i + c_j h, \mathbf{y}_i + \sum_{l=1}^{j-1} a_{jl} K_l\right). \tag{3.24}$$

The simplest case is to take $\kappa = 1, w_1 = 1$ and $\mathbf{K}_1 = h\mathbf{f}(t_i, \mathbf{y}_i)$. This gives the Euler method, which is a first-order numerical procedure for solving ODEs with a given initial value. The Euler method is the most basic explicit method for numerical integration of ODEs and is the simplest Runge-Kutta method.

To demonstrate how the Euler method is employed for numerically solving the initial value problem in ODEs, we use only one differential equation, Equation (3.15) for cell growth, as an example. We have the following iterations to generate numerical solutions step by step:

$$\begin{cases} X_0 = X(0), \\ X_{i+1} = X_i + h\mu_m(1 - \frac{X_i}{X_{\text{Max}}})X_i, i = 0, 1, \cdots, N-1 \end{cases} \tag{3.25}$$

The most commonly used higher-order and one-step numerical method for numerically solving ODEs with some initial conditions is the classical fourth-order Runge-Kutta method, where $\kappa = 4$. The numerical solution can be calculated by

$$\mathbf{y}_{i+1} = \mathbf{y}_i + \frac{1}{6}(\mathbf{K}_1 + 2\mathbf{K}_2 + 2\mathbf{K}_3 + \mathbf{K}_4) \tag{3.26}$$

$$\mathbf{K}_1 = h\mathbf{f}(t_i, \mathbf{y}_i) \tag{3.27}$$

$$\mathbf{K}_2 = \mathbf{f}\left(t_i + \frac{1}{2}h, \mathbf{y}_i + \frac{1}{2}\mathbf{K}_1\right) \tag{3.28}$$

$$\mathbf{K}_3 = \mathbf{f}\left(t_i + \frac{1}{2}h, \mathbf{y}_i + \frac{1}{2}\mathbf{K}_2\right) \tag{3.29}$$

$$\mathbf{K}_4 = \mathbf{f}(t_i + h, \mathbf{y}_i + \mathbf{K}_3) \tag{3.30}$$

for $i = 0, 1, \cdots, N-1$. Therefore, for ODE (3.15), we have

$$X_0 = X(0) \tag{3.31}$$

$$\mathbf{K}_1 = h\mu_m\left(1 - \frac{X_i}{X_{\text{Max}}}\right)X_i \tag{3.32}$$

$$\mathbf{K}_2 = h\mu_m\left(1 - \frac{X_i + \mathbf{K}_1/2}{X_{\text{Max}}}\right)\left(X_i + \frac{\mathbf{K}_1}{2}\right) \tag{3.33}$$

$$\mathbf{K}_3 = h\mu_m\left(1 - \frac{X_i + \mathbf{K}_2/2}{X_{\text{Max}}}\right)\left(X_i + \frac{\mathbf{K}_2}{2}\right) \tag{3.34}$$

$$\mathbf{K}_4 = h\mu_m\left(1 - \frac{X_i + \mathbf{K}_3}{X_{\text{Max}}}\right)(X_i + \mathbf{K}_3) \tag{3.35}$$

$$X_{i+1} = X_i + \frac{1}{6}(\mathbf{K}_1 + 2\mathbf{K}_2 + 2\mathbf{K}_3 + \mathbf{K}_4) \tag{3.36}$$

for $i = 0, 1, \cdots, N-1$.

Using this numerical scheme and starting from $t_0 = 0$ and given initial conditions $\mathbf{y}_0 = [X(0), P(0), S(0)]^T$, the numerical computation for the solution of the ODE model can be calculated from Equation (3.23) step by step till the end of the fermentation operation.

3.5 Simulation for Batch Fermentation

Let us simulate a batch fermentation using the kinetic parameters recommended by Klasson *et al.*, with the initial conditions of $X(0) = 0.1g/L, P(0) = 60g/L$ and $P(0) = 0$ for a time period of 25 hours starting from 0. The other system parameters are set as $\mu_m = 0.372, \alpha = 0.135, \beta = 0.0062, C = 1.03, D = 0.213$, and $X_{\text{Max}} = 30g/L$.

Analytical and numerical solutions to the biomass concentration $P(t)$ at discrete-time points over 25 hours are given in Table 3.1. For numercial computation, both the first- and fourth-order Runge-Kutta methods are used with $N = 25$. Again, the first-order Runge-Kutta method is also known as the Euler method. The Euler method is also tested in our simulations for $N = 250$.

It is seen from Table 3.1 that the forth-order Runga-Kutta method can approximate the solution very well. In comparison with the Euler method, the fourth-order Runge-Kutta method shows better computational accuracy even with a larger step size. However, it is understandable that the fourth-order Runge-Kutta method demands more computing power and resources.

Table 3.1 Analytical and numerical solutions to a batch fermentation process model.

Fermentation Time (Hour)	Analytical Solution	4th Runge-Kutta (N=25)	Euler Method (N=25)	Euler Method (N=250)
0	0.1	0.1	0.1	0.1
2	0.2097	0.2096	0.1878	0.2069
4	0.4378	0.4378	0.3522	0.4265
6	0.9067	0.9065	0.6579	0.8726
8	1.8464	1.8458	1.2211	1.7581
10	3.6381	3.6369	2.2392	3.4407
12	6.7517	6.7495	4.0173	6.3796
14	11.380	11.376	6.9377	10.830
16	16.877	16.874	11.254	16.284
18	21.906	21.903	16.644	21.447
20	25.519	25.517	21.950	25.252
22	27.689	27.688	25.895	27.564
24	28.856	28.855	28.154	28.805

To better visualize the difference in computing accuracy, Figure 3.2 illustrates the numerical solutions from our simulations from the fourth-order Runga-Kutta method and the Euler method.

3.6 Simulation for Fed-Batch Fermentation

Fed-batch operation of fermentation processes has advantages in improving yield and productivity. The feed profile is either to be predetermined or determined adaptively according to the maximization of a profit function.

The fed-batch fermentation we are investigating is designed with two feeding patterns: pulse feeding and step feeding. These two feeding patters are explained below:

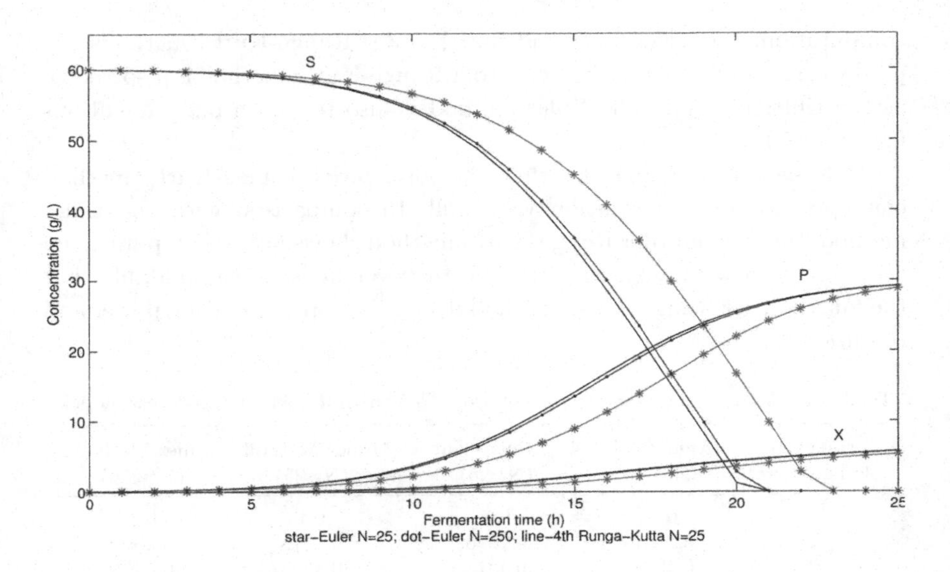

Fig. 3.2 Numerical solutions to a batch fermentation model.

- Pattern 1: pulse feeding. The addition starts at 120g/L residual glucose concentration and repeated every 24 hours. The amount of feeding is based on a dilution rate of F/V= 0.1.
- Pattern 2: step feeding. It starts at 24 hours, and then is on/off at 24 hours interval with a constant flow rate of 18mL/h.

Our simulation studies in this section use the rate equations we proposed in [Yao *et al.* (2001)]. The mathematical model of the batch fermentation process can be expressed as follows

$$\frac{dX(t)}{dt} = kS^n(t)X(t) - X(t)\frac{F(t)}{V(t)}, \tag{3.37}$$

$$\frac{dP(t)}{dt} = \alpha X(t)\log(\beta S(t) + \gamma) - P(t)\frac{F(t)}{V(t)}, \tag{3.38}$$

$$\frac{dS(t)}{dt} = -\frac{1}{Y_{x/s}}\frac{dX(t)}{dt} - \frac{1}{Y_{p/s}}\frac{dP(t)}{dt} - m_s X(t) + [S_f - S(t)]\frac{F(t)}{V(t)}, \tag{3.39}$$

$$\frac{dV(t)}{dt} = F(t). \tag{3.40}$$

The simulation uses 0.4g/L and 150g/L as initial biomass and glucose concentration. The initial working volume for batch is 2.2L and the glucose concentration in the feed stream is 400g/L. Therefore, we have the following initial conditions:

$$X(0) = 0.4g/L, S(0) = 150g/L, P(0) = 0, V(0) = 2.2L, S_f = 400g/L.$$

The Kinetic parameters of the batch fermentation process are set as:

$$k = 3.3E - 07, n = 2.5776, \alpha = 0.0754, \beta = 0.0177,$$
$$\gamma = 1.7031, Y_{x/s} = 0.2579, Y_{p/s} = 0.4537, m_s = 0.00412.$$

The batch fermentation is considered to be completed when the residual glucose reaches 0.1g/L.

The fourth-order Runge-Kutta method is used to numerically solve the fed-batch fermentation process model. According to the designed feeding pattern, the system model will be switched between batch and fed-batch operations during the running of the simulation program. In fact, for the pulse-feeding pattern, the batch model can be used throughout the fermentation process. However, for every time of feeding, the initial concentrations used for model computation should be adjusted by a factor of (1+F/V).

Numerical solutions for fed-batch fermentation with pulse feeding pattern are depicted in Figures 3.3, while simulation results for step feeding pattern is graphically illustrated in 3.4.

These simulation results in Figures 3.3 and 3.4 show that pulse feeding and step feeding lead to different process dynamics. In comparison with pulse feeding, step feeding leads to longer time for the fermentation process to finish. The concentration of the final product is higher when the fermentation process is operated with step feeding. But at the point of $t = 90$ hours, around which the fermentation process with pulse feeding finishes, both pulse feeding and step feeding operations have similar lysine concentration.

Finally, let us compare the simulation results between batch and fed-batch operations. Results from batch fermentation are previously shown in Figure 3.2. Clearly, Figures 3.3 and 3.4 for bed-batch fermentation show that repeated addition of glucose is beneficial for improved productivity and yield.

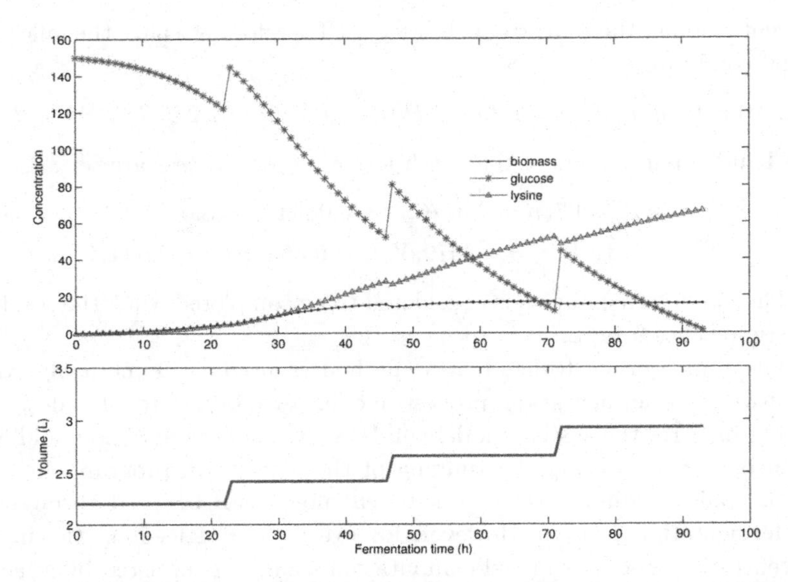

Fig. 3.3 Lysine fed-batch fermentation: The addition of glucose is chosen as pulse mode, started at 120g/L residual glucose and repeated every 24 hours.

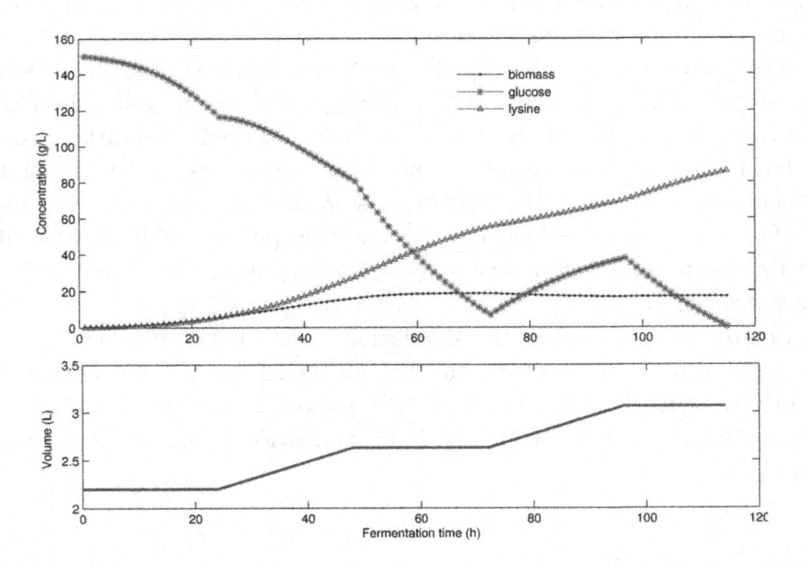

Fig. 3.4 Lysine fed-batch fermentation: The addition of glucose is in step mode, started at 24 hours and on/off every 24 hours.

Chapter 4

Finite Difference Methods for Partial Differential Equation Models

This chapter aims to develop finite difference methods for numerical computation of partial differential equation (PDE) models for complex industrial systems. We consider a continuous galvanizing process as an example for process modelling and numerical model computation. The process modelling gives a typical boundary value problem of PDEs. Finite difference methods are then developed to solve the PDEs with the boundary conditions.

This chapter is organized as follows. Section 4.1 introduces continuous galvanizing processes. Section 4.2 mathematically describes a galvanizing process using PDEs from the first principles. In Section 4.3, the fundamental relations in the PDE model are then transformed into a discrete-time and discrete-space model using the finite difference technique. This is followed by stability analysis and parameter estimation in Sections 4.4 and 4.5, respectively, for the developed model. After that, Sections 4.6 and 4.7 are devoted algorithm design and implementation, respectively. Finally, simulation studies and industrial applications are discussed in Section 4.8, which demonstrates the effectiveness of the developed process modelling and model computation methods.

Unless otherwise cited explicitly, the materials presented in this chapter are taken from our publications [Tian *et al.* (2000, 2004)] and related preliminary studies.

4.1 Continuous Galvanizing Processes

During the last three decades, there has been an increasing demand for high-quality galvanized steel products, resulting in an expansion in galvanizing facilities. The hot dip galvanizing technology has been improved to meet this demand. For widely used continuous galvanizing annealing, an effective monitoring and control system is critical in ensuring high-quality products. On-line and real-time computation and prediction of strip temperature are vital for such a system. Using a hot dip galvanizing annealing process as an example, the case studies of boundary value problems in PDEs to be discussed in this section aim to demonstrate how finite difference methods can be applied to modelling and computation of industrial processes.

The hot dip galvanizing process is a continuous hot dip galvanizing production line in an cold rolling mill. It was designed to produce 350 kilotons of products annually with strip thickness of 0.3 - 3mm. The maximum strip velocity was designed to be 183m/min. The continuous annealing furnace in the process is about 42m high and 38m long. It consists of four sections: F1 flaming preheating, F2 radiation heating, F3 cooling and C1 gas-jet cooling. The equivalent length of the strip in the furnace is 183m. Figure 4.1 shows the schematic diagram of the continuous annealing furnaces.

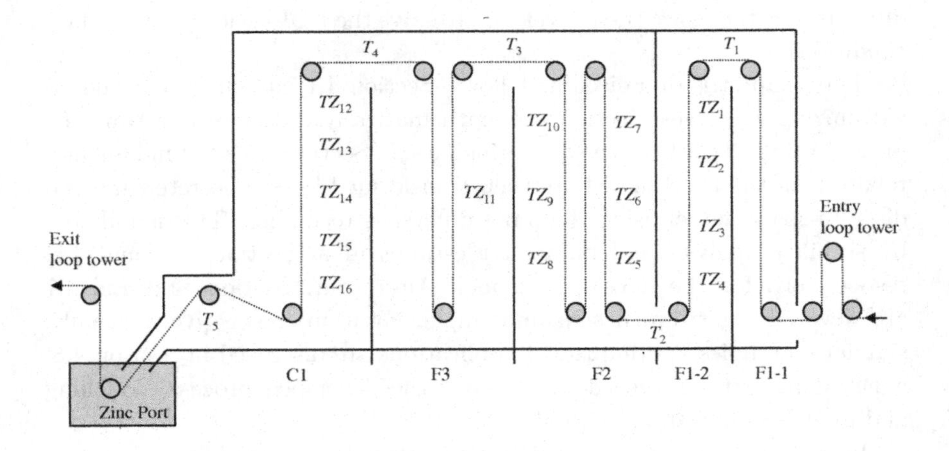

Fig. 4.1 Continuous annealing furnace in a galvanizing process.

A steel strip passes continuously through the four consequent furnace sections. It is first preheated to about 220°C in F1-1 by exhaust furnace gases and then heated to 454 - 649°C in F1-2 by a burning flame. Subsequently passing through F2, the strip is further radiantly heated to its maximum temperature by the burning of fuel and air in radiant tubes. A cooling radiant tube in F3 slowly cools the strip by passing cooled air through the tube. Finally, the strip is quickly cooled by a direct cooled gas-jet in C1 to a temperature suitable for zinc coating.

Each of the four furnace sections is further divided into one or more furnace zones. There are 16 zones altogether with corresponding furnace temperature measurements $TZ_1 - TZ_{16}$. Infrared pyrometers are installed to measure exit strip temperatures ($T_2 - T_5$) of the four sections. The exit strip temperature of F1-1 is also measured (T_1).

As strip temperatures $T_2 - T_5$ determine the quality of the products, maintaining these temperatures at specified values is crucial in the galvanizing production. A typical strip temperature profile is tabulated in Table 4.1. These strip temperatures $T_1 - T_5$ are controlled through manipulating furnace temperatures TZ_1 through TZ_{16}, while the furnace temperatures are regulated through combustion control. Corresponding to the strip temperature profile in Table 4.1, a typical furnace temperature profile is depicted in Table 4.2 and Figure 4.2.

Table 4.1 A typical strip temperature profile (°C).

T_1	T_2	T_3	T_4	T_5
450	650	760	630	465

Table 4.2 A typical furnace temperature profile (°C) corresponding to Table 4.1.

TZ_1	TZ_2	TZ_3	TZ_4	TZ_5	TZ_6	TZ_7	TZ_8	TZ_9	TZ_{10}
1127	1131	1181	1097	882	847	764	895	912	888
TZ_{11}	TZ_{12}	TZ_{13}	TZ_{14}	TZ_{15}	TZ_{16}				
210	507	413	205	195	182				

Therefore, a cascade control is adopted for strip temperatures control as shown in Figure 4.3, where TC, SD, and TZC_1–TZC_4 represent strip temperature controller, control signal distributor, and furnace temperature controllers, respectively. The inner loop of the control system is the combustion control for furnace temperatures, and the otter loop is for strip temperature control.

The design of the strip temperature controllers requires accurate estimation and prediction of the strip temperature distribution,

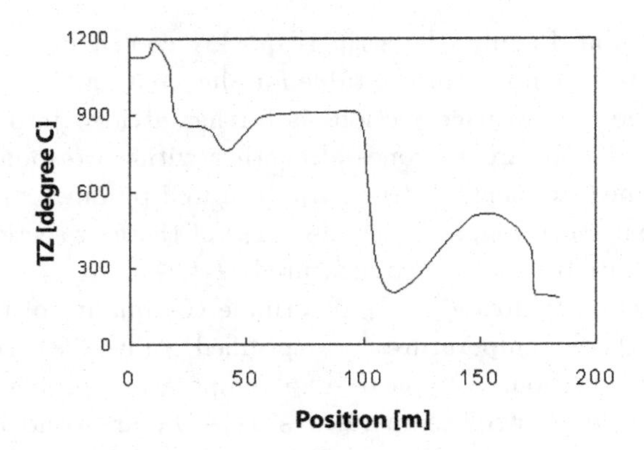

Fig. 4.2 A typical furnace temperature profile.

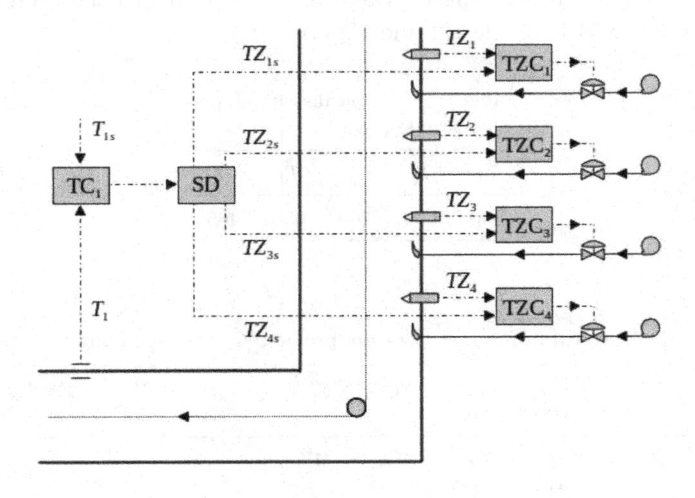

Fig. 4.3 Cascade control for furnace section F1.

necessitating modelling and model computation of the strip temperature distribution.

It is worth mentioning that the operation and control of a galvanizing annealing furnace are different from those of other continuous annealing furnaces in that a constant strip velocity is required to ensure the zinc coating quality. Thus, the strip velocity cannot be used as a manipulated variable to control the strip temperature, hence complicating the control of the galvanizing line.

4.2 Development of a PDE Model

To describe the strip temperature distribution mathematically in the continuous galvanizing annealing furnace, the following assumptions are made:

1) The movement of the steel strip in the furnace is considered equivalent along the horizontal direction in which the strip enters and exits from the furnace;
2) No temperature gradient exists in the strip width direction;
3) The furnace temperature is symmetrical over the top and bottom of the strip;
4) All other conditions for the strip top and bottom are the same; and
5) The average strip temperature across the strip thickness is computed from PDEs of heat transfer and used as a state variable.

To describe the strip temperature distribution T in the furnace, a fixed two-dimensional Cartesian coordinate system is taken with x and y representing the directions across the strip thickness and length, respectively. The origins of x and y are chosen to be at the strip bottom and the furnace entry point, respectively, as shown in Figure 4.4.

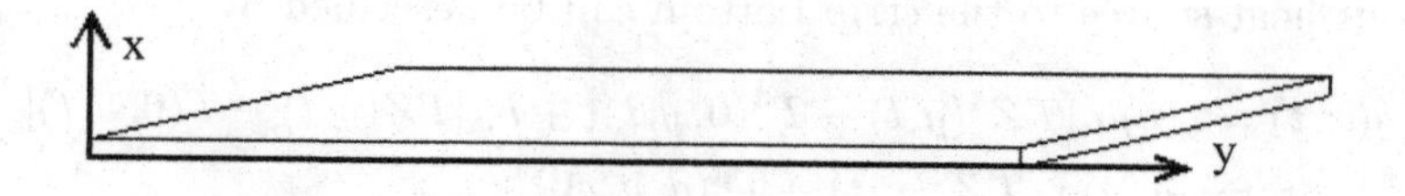

Fig. 4.4 Fixed two-dimensional coordinate system.

The furnace section F1-2 is modelled here as an example. Similar principles can be applied to other sections. Heat transfer laws are used to derive fundamental equations to describe the strip temperature distribution. The principal means of heat transfer in the strip is heat conduction. According to the Fourier law of heat conduction [10], a two-dimensional strip temperature distribution $T(x,y,t)$ for unsteady heat conduction is expressed by:

$$\frac{\partial T(x,y,t)}{\partial t} = \frac{1}{C\rho}\left\{\frac{\partial}{\partial x}\left[K_s\frac{\partial T(x,y,t)}{\partial x}\right] + \frac{\partial}{\partial y}\left[K_s\frac{\partial T(x,y,t)}{\partial y}\right]\right\} - v(t)\frac{\partial T(x,y,t)}{\partial y}, \quad (4.1)$$

with the initial condition

$$T(x,y,0) = T0(x,y), \tag{4.2}$$

where $0 \leq x \leq d(y), 0 \leq y \leq L$ and $t \geq 0$, $v(t)$ is the strip speed in m/s.

If K_s is considered to be constant, i.e. temperature and time independent, Equation (4.1) can be simplified to:

$$\frac{\partial T(x,y,t)}{\partial t} = \frac{K_s}{C\rho}\left\{\frac{\partial^2 T(x,y,t)}{\partial x^2} + \frac{\partial^2 T(x,y,t)}{\partial y^2}\right\} - v(t)\frac{\partial T(x,y,t)}{\partial y}. \tag{4.3}$$

According to the modeling assumptions, the heat flux $q(y,t)$ from the gaseous heat source to the strip bottom surface is the same as that to the strip top surface. The heat flux $q(y,t)$ is:

$$q(y,t) = -K_s \left.\frac{\partial T(x,y,t)}{\partial x}\right|_{x=0} = K_s \left.\frac{\partial T(x,y,t)}{\partial x}\right|_{x=d}. \tag{4.4}$$

It is noted that $q(y,t)$ can also be obtained directly from the heat transfer relation between the gaseous heat source and the strip surfaces. For furnace section F1-2, heat is provided primarily by combustion. It is transferred from furnace gases to the strip surfaces by both radiation and convection, with the radiation as the dominant process. The heat flux $q(y,t)$ from the gaseous heat source to the strip bottom can be described by:

$$\begin{aligned} q(x,t) &= \epsilon(y)\sigma\left[TZ^4(y,t) - T^4(0,y,t)\right] + h_c\left[TZ(y,t) - T(0,y,t)\right] \\ &= -\epsilon(y)\sigma\left[TZ^4(y,t) - T^4(d,y,t)\right] \\ &\quad -h_c\left[TZ(y,t) - T(d,y,t)\right], \end{aligned} \tag{4.5}$$

where the first and second terms correspond to radiation and convection, respectively.

Combining Equations (4.4) and (4.5) gives the following two boundary conditions:

$$\begin{aligned} \left.\frac{\partial T(x,y,t)}{\partial x}\right|_{x=0} &= -\frac{\epsilon(y)\sigma}{K_s}\left[TZ^4(y,t) - T^4(0,y,t)\right] \\ &\quad -\frac{h_c}{K_s}\left[TZ(y,t) - T(0,y,t)\right], \end{aligned} \tag{4.6}$$

$$\begin{aligned} \left.\frac{\partial T(x,y,t)}{\partial x}\right|_{x=d} &= \frac{\epsilon(y)\sigma}{K_s}\left[TZ^4(y,t) - T^4(d,y,t)\right] \\ &\quad +\frac{h_c}{K_s}\left[TZ(y,t) - T(d,y,t)\right]. \end{aligned} \tag{4.7}$$

According to Assumption (5), the average strip temperature across the strip thickness, $TS(y,t)$, is obtained through integrating $T(x,y,t)$ with respect to x from 0 to d and then being divided by d, i.e.,

$$TS(y,t) = \frac{1}{d}\int_0^d T(x,y,t)dx. \tag{4.8}$$

PDE (4.1) or (4.3) together with Equations (4.2) and (4.6) through (4.8) form fundamental relations for strip temperature distribution $T(x,y,t)$. Due to the unavailability of an analytical solution, the above relations cannot be directly used for real-time application.

The first effort to simplify these obtained relations is to linearize the boundary conditions in Equations (4.6) and (4.7). Define $h(y,t)$ as an equivalent heat transfer coefficient between the furnace and the bottom surface of the strip:

$$h(y,t) = \epsilon(y)\sigma\left[TZ^2(y,t) + T^2(0,y,t)\right]\left[TZ(y,t) + T(0,y,t)\right] + h_c. \tag{4.9}$$

Then, the boundary conditions in Equations (4.6) and (4.7) can be respectively rewritten as

$$\left.\frac{\partial T(x,y,t)}{\partial x}\right|_{x=0} = -\frac{h(y,t)}{K_s}\left[TZ(y,t) - T(0,y,t)\right], \tag{4.10}$$

$$\left.\frac{\partial T(x,y,t)}{\partial x}\right|_{x=d} = \frac{h(y,t)}{K_s}\left[TZ(y,t) - T(d,y,t)\right]. \tag{4.11}$$

"Linearized" boundary conditions (4.10) and (4.11) are simple expressions suitable for real-time applications. The equivalent heat transfer coefficient $h(y,t)$ is determined on-line through system identification, which will be discussed later.

The engineering implementation of the simplified fundamental heat transfer relations will be considered in the following sections.

4.3 Discrete State Space Model

The fundamental PDE model developed from the first principle through investigating heat transfer within the steel strip and between the furnace and strip are difficult to solve analytically. Numerical methods are commonly used to solve such a problem. In numerical computation, the fundamental relations are converted to coupled iteration equations. Among various numerical computation methods, the finite difference technique is adopted here in order to develop a simple and practical model suitable for real-time applications. The derived model is a discrete state space model, which is discrete in both time and space.

First of all, partition x into N_x sections of length Δx, and y into N_y sections of length Δy, respectively, where

$$N_x \times \Delta x = d, \ N_y \times \Delta y = L. \tag{4.12}$$

This results in $N_x \times N_y$ parallelograms on the xy space, as shown in Figure 4.5. For simplicity, let i and j denote $i\Delta x$ and $j\Delta y$, respectively, unless otherwise specified explicitly.

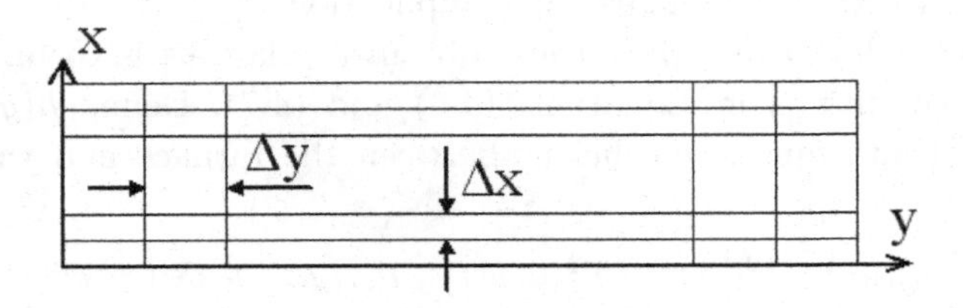

Fig. 4.5 Space partitioning.

Secondly, discretize the time t into a finite number of time instants with time step of Δt. For simplicity, let k denote $k\Delta t$.

Assume that K_s remains constant and no gradients exist for all state and manipulated variables within the same parallelogram and in the same time interval. Then, when the finite difference technique is applied, the Euler method is adopted to simplify the problem solving. Also, both forward and backward differences will be utilized in order to derive a mathematical model that is relatively easy to handle. Specifically, the first-order term $\frac{\partial T(x,y,t)}{\partial t}$ is approximated by forward difference; while the two first-order terms $\frac{\partial T(x,y,t)}{\partial y}$ and $\frac{\partial T(x,y,t)}{\partial y}$ are approximated by backward difference, i.e.,

$$\frac{\partial T(x,y,t)}{\partial t} \approx \frac{T(i,j,k+1) - T(i,j,k)}{\Delta t}, \tag{4.13}$$

$$\frac{\partial T(x,y,t)}{\partial x} \approx \frac{T(i,j,k) - T(i-1,j,k)}{\Delta x}, \tag{4.14}$$

$$\frac{\partial T(x,y,t)}{\partial y} \approx \frac{T(i,j,k) - T(i,j-1,k)}{\Delta y}. \tag{4.15}$$

Furthermore, both the two second-order terms $\frac{\partial T(x,y,t)}{\partial t}$ and $\frac{\partial T(x,y,t)}{\partial y}$

are approximated by mixed forward-backward differences, i.e.,

$$\frac{\partial^2 T(x,y,t)}{\partial x^2} \approx \frac{T(i+1,j,k) - 2T(i,j,k) + T(i-1,j,k)}{\Delta x^2}, \tag{4.16}$$

$$\frac{\partial^2 T(x,y,t)}{\partial y^2} \approx \frac{T(i,j+1,k) - 2T(i,j,k) + T(i,j-1,k)}{\Delta y^2}. \tag{4.17}$$

Substituting Equations (4.13) and (4.15) through (4.17) into PDE (4.3) yields:

$$\begin{aligned} T(i,j,k+1) &= aT(i+1,j,k) + (1-2a-2b-c)T(i,j,k) \\ &\quad +aT(i-1,j,k) + bT(i,j+1,k) + (b+c)T(i,j-1,k), \\ &\quad 1 \le i \le N_x - 1, \ \ 1 \le j \le N_y - 1, \end{aligned} \tag{4.18}$$

where

$$a = K_s \frac{\Delta t}{C\rho \Delta x^2}, \ b = K_s \frac{\Delta t}{C\rho \Delta y^2}, \ c = v(k)\frac{\Delta t}{\Delta y}. \tag{4.19}$$

Then, substituting Equation (4.14) into linearized boundary conditions (4.10) and (4.11) gives the following two discretized boundary conditions:

$$T(1,j,k) \approx T(0,j,k) - \frac{h(j,k)\Delta x}{K_s}\left[TZ(j,k) - T(0,j,k)\right], \tag{4.20}$$

$$\begin{aligned} T(N_x,j,k) &\approx T(N_x-1,j,k) \\ &\quad -\frac{h(j,k)\Delta x}{K_s}\left[TZ(j,k) - T(N_x,j,k)\right]. \end{aligned} \tag{4.21}$$

With the strip temperature distribution expressed in Equations (4.18) through (4.21), the average strip temperature $TS(y,t)$ shown in Equation (4.8) can be approximated by:

$$\begin{aligned} TS(j,k) &\approx \lim_{N_x\to\infty} \frac{1}{d}\sum_{i=1}^{N_x} T(i,j,k)\Delta x \\ &= \lim_{N_x\to\infty} \frac{1}{N_x}\sum_{i=1}^{N_x} T(i,j,k), \ \ 0 \le j \le N_y. \end{aligned} \tag{4.22}$$

Therefore, taking the sum for i on both sides of Equation (4.18) for $1 \le j \le N_y - 1$ and simplifying the results yield:

$$\begin{aligned} TS(j,k+1) &\approx (1-2b-c)TS(j,k) + (b+c)TS(j-1,k) + bTS(j+1,k) \\ &\quad + \lim_{N_x\to\infty}\frac{1}{N_x}[T(N_x,j,k) - T(N_x-1,j,k) + T(0,j,k) \\ &\quad -T(1,j,k)], \ 1 \le j \le N_y - 1. \end{aligned} \tag{4.23}$$

Considering Equations (4.20) and (4.21), we can rewrite Equation (4.23) as:

$$\begin{aligned}TS(j,k+1) \approx{}& (1-2b-c-d_{jk})TS(j,k)+(b+c)TS(j-1,k)\\&+bTS(j+1,k)+d_{jk}TZ(j,k), 1\le j\le N_y-1,\end{aligned} \quad (4.24)$$

where

$$d_{jk}=\frac{2h(j,k)\Delta t}{C\rho d}. \quad (4.25)$$

Similarly, for $j=0$ and $j-N_y$, we have the following two approximations:

$$TS(0,k+1) \approx (1-b-d_{0k})TS(0,k)+(b+c)TS(1,k)+d_{0k}TZ(0,k), \quad (4.26)$$

$$\begin{aligned}TS(N_y,k+1) \approx{}& (b+c)TS(N_y-1,k)+(1-b-c-d_{N_yk})TS(N_y,k)\\&+d_{N_yk}TZ(N_y,k).\end{aligned} \quad (4.27)$$

Equations (4.24) through (4.27) form a scalar discrete-time and discrete-space state space model for average strip temperature distribution along the strip length direction.

In order to use linear system theory for system identification and control, the scalar state space model in Equations (4.24) through (4.27) is converted to a matrix representation. For this purpose, denote

$$X(k)=\begin{bmatrix}TS(0,k)\\TS(1,k)\\ \vdots\\TS(N_y,k)\end{bmatrix}, \quad U(k)=\begin{bmatrix}TZ(0,k)\\TZ(1,k)\\ \vdots\\TZ(N_y,k)\end{bmatrix}, \quad (4.28)$$

$$A(k)=\begin{bmatrix}a_{1,1} & b & & & \\ b+c & a_{2,2} & b & & \\ & \ddots & \ddots & \ddots & \\ & & b+c & a_{N_y,N_y} & b\\ & & & b+c & a_{N_y+1,N_y+1}\end{bmatrix}, \quad (4.29)$$

$$B(k)=diag\{d_{0k},d_{1k},\cdots,d_{N_yk}\}, \quad (4.30)$$

where X and U represent state and control vectors, respectively;

$$\begin{cases}a_{1,1}=1-b-d_{0k},\\ a_{p,p}=1-2b-c-d_{p-1,k}, \quad p=2,3,\cdots,N_y,\\ a_{N_y+1,N_y+1}=1-b-c-d_{N_yk}.\end{cases} \quad (4.31)$$

Obviously, matrix A is a sparse matrix with most of its elements being zero, and matrix B is a diagonal matrix. The non-zero elements of A and B need to be estimated on-line.

With the notations in Equations (4.28) through (4.31), the scalar state space model in Equations (4.24) through (4.27) can be expressed in a compact matrix representation as follows:

$$X(k+1) = A(k)X(k) + B(k)U(k). \tag{4.32}$$

This matrix state space model will be further used for system identification and control design.

4.4 Stability Analysis and Parameter Settings of the Model

The developed discrete state space model in Equation (4.32) together with Equations (4.28) through (4.31) involves a number of parameters representing the physical properties and operating conditions of the galvanizing process, such as $C, \rho, d, v(t)$, etc. These parameters are not adjustable in model computation unless the operating environment has changed, e.g., when a new strip roll arrives. In addition to these parameters, there are two adjustable parameters, time step size Δt and space step size Δy, in model computation.

The two adjustable parameters Δt and Δy determine the computational complexity, accuracy and stability of the model, and thus need to be chosen carefully. Bigger values of Δt and Δy allow reduced computational demand while giving less accurate results or leading to unstable model computation. In contrast, smaller values of Δt and Δy benefit the computational accuracy and stability, but will result in increased demand on computational power and tighter requirements on real-time computation. Therefore, a compromise has to be made when selecting values of Δt and Δy in model implementation.

Constraints on the adjustable parameters Δt and Δy may be obtained through model stability analysis. The state space model in Equation (4.32) together with Equations (4.28) through (4.31) is stable if the absolute maximum eigenvalue of $A(k)$ is less than 1. Since any arbitrary eigenvalue λ of $A(k)$ satisfies

$$|\lambda| \leq \max_j \sum_{i=1}^{n} |a_{ij}|, \tag{4.33}$$

it follows that the model is stable if $\max_j \sum_{i=1}^{n} |a_{ij}| < 1$. From the definition of matrix A, the following stability conditions can be obtained, which should hold true simultaneously:

$$\begin{cases} |1 + c - d_{0k}| < 1, \\ |1 - d_{jk}| < 1, \quad 1 \le j \le N_y - 1, \\ |1 - c - d_{N_y k}| < 1. \end{cases} \tag{4.34}$$

These conditions lead to the following inequalities:

$$\begin{cases} c < \ d_{0k} \ < 2 + c, \\ 0 < \ d_{jk} \ < 2, \qquad 1 \le j \le N_y - 1, \\ -c < d_{N_y k} < 2 - c. \end{cases} \tag{4.35}$$

Taking into account the definitions of c and d_{jk} in Equations (4.19) and (4.25), respectively, we can now derive from Equation (4.35) the stability conditions of the model computation for $\Delta t > 0$ and $\Delta y > 0$ as:

$$\begin{cases} v(k)\dfrac{\Delta t}{\Delta y} < 2\dfrac{h(0,k)\Delta t}{C\rho d} < 2 + v(k)\dfrac{\Delta t}{\Delta y}, \\ h(j,k)\Delta t < C\rho d, \ 1 \le j \le N_y - 1, \\ 2h(N_y,k)\Delta t < \left[2 - v(k)\dfrac{\Delta t}{\Delta y}\right] C\rho d. \end{cases} \tag{4.36}$$

The constraints in Equation (4.36) are used to guide the selection of the values of the adjustable parameters Δt and Δy for model implementation. It can be seen from Equation (4.36) that large values of Δt and Δy will result in dissatisfaction of the stability conditions. However, small values of Δt and Δy increase computation and storage requirements as mentioned previously. For the process considered in this example, setting $\Delta t = 4$s and $\Delta y = 10$m is a satisfactory choice for which the model is stable and the computation and storage requirements are also acceptable.

4.5 Identification of Model Parameters

Model parameters in system matrices $A(k)$ and $B(k)$ depend on the time and space step sizes, the characteristics of the furnace and steel strip, and the operating conditions such as strip velocity and the desired strip temperatures. These parameters are thus required to be identified on-line and in real time for industrial applications. This section discusses specific techniques used in this example for parameter identification.

First of all, it is necessary to clarify how many and which parameters need to be identified. The definitions of the system matrices $A(k)$ and $B(k)$ indicate that the following list of parameters must be determined:

$$C, \rho, K_s, d, v(k), \Delta t, \Delta y, h(j,k).$$

Parameters C, ρ, K_s, d can be determined in advance before the steel strip moves into the furnace. For example, for a typical steel roll, we have $C = 465, \rho = 7833\text{kg/m}^3, K_s = 42, d = 0.008\text{m}$. The operating strip velocity $v(k)$ is measured on-line and in real time, and a typical setting of v is $v = 120\text{m/min} = 2\text{m/s}$. The time step Δt and space step Δy have been determined to be 4s and 10m, respectively, through model stability analysis in the last section. In this case, a, b and c can be computed directly. In particular, b is in the order of 10^{-8} and thus is negligible in comparison with other parameters, implying that the matrix $A(k)$ in Equation (4.29) can be approximated by:

$$A(k) \approx \begin{bmatrix} a_{1,1} & b & & & \\ b+c & a_{2,2} & b & & \\ & \ddots & & \ddots & \\ & b+c & a_{N_y,N_y} & b & \\ & & b+c & a_{N_y+1,N_y+1} \end{bmatrix}, \tag{4.37}$$

where

$$\begin{cases} a_{1,1} \approx 1 - d_{0k}, \\ a_{p,p} \approx 1 - c - d_{p-1,k}, \quad p = 2, 3, \cdots, N_y, \\ a_{N_y+1,N_y+1} \approx 1 - c - d_{N_y k}. \end{cases} \tag{4.38}$$

Such approximations further reduce the complexity of the model identification. Therefore, only the equivalent heat transfer coefficient $h(j, k)$ has to be identified on-line and in real time.

The definition of $h(y, t)$ in Equation (4.9) shows that h depends on furnace temperature TZ, strip temperature T, equivalent darkness coefficient ϵ and convection heat transfer coefficient h_c. From the established discrete state space model in Equations (4.28) through (4.32), an identification program is developed to compute $h(j, k)$ on-line and in real time, which employs standard recursive least-square algorithm with a properly selected forgetting factor. Figure 4.6 depicts a typical plot of h versus y identified from actual measurements of furnace and strip temperatures under $\Delta t = 4\text{s}$ and $\Delta y = 10\text{m}$. It is seen from Figure 4.6 that $h(j, k)$ varies with the changes in time k and strip position j. The introduction of h into the model can reduce the computational demand of the process model effectively.

In order to apply linear system theory to system identification of the matrix state space model in Equations (4.28) through (4.32), instead of identifying h, the matrices A and B can be identified on-line as an

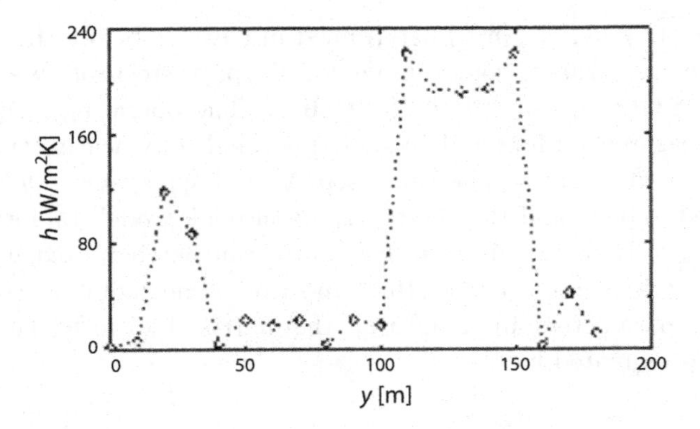

Fig. 4.6 A plot of h versus y at a specific time instant.

alternative method for model identification. As mentioned previously, A is sparse and B is diagonal. Only non-zero elements of A and B need to be identified. As a result, the computational requirement is significantly reduced. This method alleviates the constraints on determining some parameters such as C, ρ, K_s, d and $v(k)$ prior to the identification, increasing the model flexibility and accuracy. Thus, it has been adopted in this example for industrial applications. The identification is implemented with the recursive least-square algorithm, which will be discussed later.

In the system under investigation, the actual measurement points for strip and furnace temperatures are only 5 and 16, respectively. The second critical problem in model parameter identification is that the number of state and control variables is more than that of actual measurements. This introduces an extra difficulty to model identification. To overcome this difficulty, unmeasured furnace temperatures are determined by interpolation. The same method is applied to unmeasured strip temperatures for initial iteration. Assume that, at time $k-1$, the predicted strip temperatures at all measurement points $j = j_i$ are $TS_p(j_i, k), i = 1, 2, \cdots, 5$. If the measured values of these strip temperatures at time k are $TS_a(j_i, k), i = 1, 2, \cdots, 5$, then for $\forall j, j_i \leq j \leq j_{i-1}$, the strip temperature $TS(j, k)$, which is required in model computation and identification at time k, is determined by:

$$TS(j,k) = \left[\frac{j_{i+1} - j}{j_{i+1} - j_i} \frac{TS_a(j_i,k)}{TS_p(j_i,k)} + \frac{j - j_i}{j_{i+1} - j_i} \frac{TS_a(j_{i+1},k}{TS_p(j_{i+1},k)}\right] TS_p(j,k),$$
$$\forall j : j_i \leq j \leq j_{i-1},\ i = 1, 2, \cdots, 5. \tag{4.39}$$

Moreover, changes in strip velocity v will lead to variations in system parameters. However, the parameter variations cannot be macroscopically identified within a short time, resulting in a time delay in both system identification and control. It is noted that only c depends on v in the form of $c(k) = v(k)\frac{\Delta t}{\Delta y}$ as defined in Equation (4.19). Assume that an estimation $\hat{c}(k-1)$ has been obtained for $c(k-1)$ at time $k-1$. If v changes at time k, then instead of the identified c(k), a new estimation $\hat{c}(k)$ can be used for model computation as shown in the following equation:

$$\hat{c}(k) = \frac{v(k)}{v(k-1)}\hat{c}(k-1) \tag{4.40}$$

This increases the adaptive ability of the process model.

4.6 Least-Square Algorithms for System Identification

The fundamental algorithms for model identification are simple. For example, the standard least-square identification algorithm and its extensions can be used. For easier discussion of the algorithm developed in this work, a brief description of the standard least-square identification algorithms is given below.

Consider a model

$$z(t) = \phi_1(t)\theta_1 + \phi_2(t)\theta_2 + \cdots + \phi_n(t)\theta_n = \phi^T(t)\Theta \tag{4.41}$$

where ϕ and θ are regressor and parameter vectors, respectively:

$$\begin{aligned} \phi(t) &= [\phi_1(t), \phi_2(t), \cdots .\phi_n(t)]^T, \\ \Theta &= [\theta_1, \theta_2, \cdots, \theta_n]^T. \end{aligned} \tag{4.42}$$

The model in Equation (4.41) is linear in the parameters.

A series of observations of z is obtained together with corresponding ϕ, $\{(z(i), \phi(i), i = 1, 2, \cdots, t\}$. Denote

$$Z(t) = \begin{bmatrix} z(1) \\ z(2) \\ \vdots \\ z(t) \end{bmatrix}, \quad \Phi(t) = \begin{bmatrix} \phi^T(1) \\ \phi^T(2) \\ \vdots \\ \phi^T(t) \end{bmatrix}. \tag{4.43}$$

It follows that

$$Z(t) = \Phi(t)\Theta. \tag{4.44}$$

Furthermore, if the estimation of $z(i)$ is denoted by $\hat{z}(i)$ and the estimation error for $z(i)$ is denoted by $e(i)$, we have

$$e(i) = z(i) - \hat{z}(i) = z(i) - \phi^T(i)\Theta, \ i = 1, \cdots, t. \tag{4.45}$$

We define estimation error vector

$$E(t) = \left[e(1), e(2), \cdots, e(t)\right]^T. \tag{4.46}$$

Then, we formulate the following minimization problem with respect to the parameter vector Θ:

$$\min_{\Theta} E^T(t)E(t) = \sum_{i=1}^{t} e(i). \tag{4.47}$$

Solving this minimization problem gives an estimation of the parameter vector Θ:

$$\begin{cases} \hat{\Theta}(t) = P(t)\Phi^T Z(t) = P(t)\sum_{i=1}^{t}\phi(i)z(i) \\ P(t) = \left(\Phi^T\Phi\right)^{-1} = \left[\sum_{i=1}^{t}\phi(i)\phi^T(i)\right]^{-1} \end{cases} \tag{4.48}$$

The solution $\hat{\Theta}$ in (4.48) requires that $\Phi^T\Phi$ be non-singular.

To recursive form of the above least-square estimation algorithm for Θ is expressed by:

$$\begin{cases} \hat{\Theta}(t) = \hat{\Theta}(t-1) + K(t)\left[z(t) - \phi^T(t)\hat{\Theta}(i-1)\right], \\ K(t) = P(t)\phi(t) = \dfrac{P(t-1)\phi(t)}{1 + \phi^T(t)P(t-1)\phi(t)}, \\ P(t) = \left[I - K(t)\phi^T(t)\right]P(t-1). \end{cases} \tag{4.49}$$

To smooth out the estimation through gradually forgetting old history, a forgetting factor λ is introduced into the estimation scheme. The corresponding estimation $\hat{\Theta}$ is

$$\begin{cases} \hat{\Theta}(t) = \hat{\Theta}(t-1) + K(t)\left[z(t) - \phi^T(t)\hat{\Theta}(i-1)\right] \\ K(t) = P(t)\phi(t) = \dfrac{P(t-1)\phi(t)}{\lambda + \phi^T(t)P(t-1)\phi(t)} \\ P(t) = \left[I - K(t)\phi^T(t)\right]P(t-1)/\lambda \end{cases} \tag{4.50}$$

where the forgetting factor $\lambda \in [0, 1]$, and usually a value of $\lambda \in [0.95, 1]$ is chosen.

4.7 Simplification of System Identification Algorithms

For the developed state space model in Equations (4.28) through (4.32), recall that Δy is determined to be 10m, implying that A is a 19×19 matrix for the given strip length of $L = 183$m in the furnace. Because the above-mentioned standard least-square algorithms are applied to each of the 19 rows of the matrix A, there will be a demand of heavy computation of matrices and matrix inverses in estimation of A. This prevents us from using the standard algorithms for real-time implementation of the state space model, necessitating more efficient algorithms through algorithm simplification for the specific process model.

It is seen from Equation (4.29) that the matrix $A(k)$ is sparse with most of its elements being zero. The first and last row of A has respective two elements, and all other rows have only three elements! This structural feature of A can be used to develop some algorithms with significant reduction of computational demand.

The following is a brief description of such an algorithm developed in this work. The basic framework of the algorithm is similar to (4.50), i.e. the least-square scheme with a forgetting factor. However, the calculation for each row of $A(k)$ is carried out over two or three elements, compared to nineteen elements of the standard algorithms! In the following descriptions of the developed algorithm, the vector Θ is formed by the two or three non-zero elements of a row of A while $\phi(t)$ is formed by the corresponding two or three elements of the state vector $X(t)$. The measurement $z(t)$, i.e. T in the model of the steel strip temperature distribution, corresponds to a specific point of y.

The following four steps form the fundamental algorithm for model identification in this work:

1) Step 1. For each row of $A(k)$, construct ϕ and θ using the two or three non-zero elements in the corresponding row of the matrix A.
2) Step 2. For each row of $A(k)$, calculate $K(t)$ based on $\phi(t)$ and $P(t-1)$. The formula is

$$K(t) = \frac{P(t-1)\phi(t)}{0.95 + \phi^T(t)P(t-1)\phi(t)}. \tag{4.51}$$

3) Step 3. For each row of $A(k)$, update P based on $K(t)$ and ϕ. The equation is

$$P(t) = \left[I - K(t)\phi^T(t)\right] P(t-1)/0.95. \tag{4.52}$$

4) Step 4. For each row of $A(k)$, identify the two or three non-zero elements that form the corresponding θ. The formula is

$$\hat{\theta}(t) = \hat{\theta}(t-1) + K(t)\left[z(t) - \phi^T(t)\hat{\theta}(i-1)\right]. \tag{4.53}$$

The developed model identification algorithm can be further simplified. This is based on the observation from the system analysis that not all non-zero elements of the matrix $A(k)$ need to be identified.

Let us consider the pth row of the matrix $A(k), 1 < p < 19$. It is seen from Equation (4.29) that the three non-zero elements in the row are

$$\begin{cases} a_{p-1,p} = b + c, \\ a_{p,p} = 1 - 2b - c - d_{p-1,k}, \\ a_{p,p+1} = b. \end{cases} \tag{4.54}$$

It is seen from Equation (4.54) that $a_{p,p+1} = b$ and $a_{p-1,p} = b + c = a_{p,p+1} + c$ for $1 < p < 19$. This means that either $a_{p-1,p}$ or $a_{p,p+1}$ can be excluded from the identification scheme as they can be derived from each other with the known value of c. It is decided to exclude $a_{p,p+1} = b$ because b is in the order of 10^{-8} and thus is actually negligible in our typical applications, as seen in the approximations in Equations (4.37) and (4.38).

Therefore, only one parameter $a_{1,1}$ needs to be identified for the first row of $A(k)$. For all other rows of $A(k)$, only two parameters, $a_{p-1,p}$ and $a_{p,p}$, are required to be identified. This implies that the problem of identifying a 19×19 matrix becomes a problem of identifying a scalar parameter and 18 additional pairs of parameters. Consequently, the complexity of the system identification is reduced significantly.

Furthermore, from Equation (4.19), $c = v\Delta t/\Delta y$ can be calculated directly because the strip velocity $v(k)$ can be measured in real time and Δt = 4s and Δy = 10m are pre-determined model parameters.

4.8 Simulations and Industrial Applications

Using the proposed modelling, a full-furnace dynamic model was developed for strip temperature distribution in the furnace. The model was implemented on a computer, which is an intelligent terminal in a computer network for real-time control and operation of the hot dip galvanizing production line. With online identification capability, it was used to predict strip temperature distribution for model-based predictive control.

With the developed model, process behaviours can be easily simulated in a wide range of operating conditions, particularly for those conditions that are not allowed to conduct experiments in real operations. For example, perturbations in strip velocity are usually due to abnormal production operations, on which industrial experiments are not permitted in real production. However, investigations into the response of the strip temperature distribution to a step change in strip velocity is helpful for understanding the dynamic behaviours of the process. This can be achieved through simulations using the developed model.

For step changes of 30m/min in strip velocity, typical simulation results are shown in Figure 4.7. The simulation results well reflect the dynamic behavious of the strip temperature in the furnace. For example, it is confirmed from the simulation studies that among T_2 to T_5, T_2 is the most sensitive to strip velocity changes.

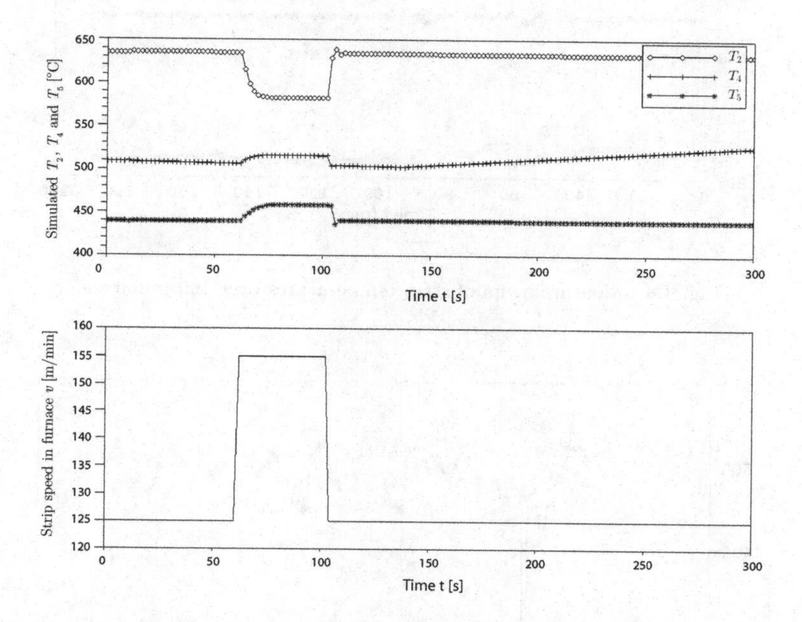

Fig. 4.7 Simulated step responses of T_2, T_4 and T_5 to step changes in strip velocity v.

The developed model has been integrated into a real-time control system over computer networks for a real industrial galvanizing line. Industrial applications of the model have shown that the model predictions of the strip temperatures match actual temperature measurements well at all measurement points. Figure 4.8 depicts actual measurements of T_1 to T_5 over three hours. It is plotted with samples taken every 15 sampling periods (4 s $\times$ 15 = 1 min) from the measured temperature data. For comparison, Figure 4.9 gives model predictions and actual measurements of T_4 for the same period of time over three hours. It is seen from this figure that the predicted T_4 values are very close to measured T_4 values.

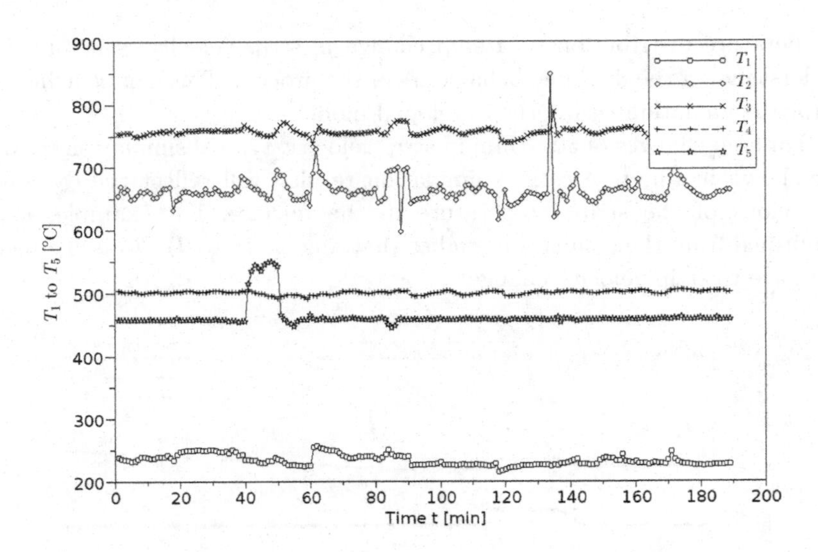

Fig. 4.8 Measurements of strip temperatures over 190 minutes.

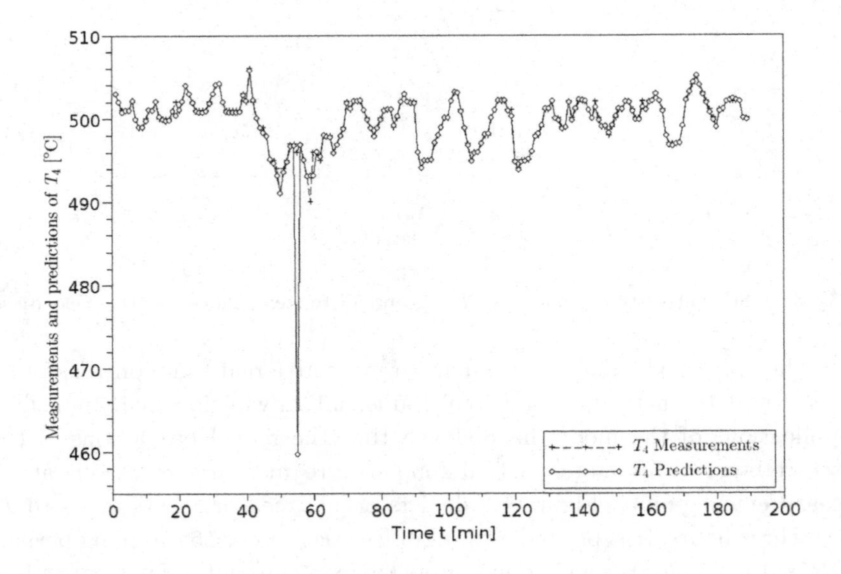

Fig. 4.9 Measurements and predictions of T_4 over 190 minutes.

To further evaluate the performance of the developed model, a detailed quantitative analysis has been conducted for predicted and measured strip temperature values. Corresponding to Figures 4.8 and 4.9, predictive errors of strip temperatures T_2 to T_4 are shown in Figure 4.10 for the same period of time of over three hours. Spikes are observed in Figure 4.10 at around $t =$ 60min for T_2 to T_5. An investigation has shown that these spikes happened at the time instant when a new strip roll entered the furnace. The rough welding seam between the new and old strip rolls may cause big errors in strip temperature measurement.

Statistical analysis is also carried out for the predictive errors of the strip temperatures T_1 to T_5 as shown in Figure 4.10 obtained in real industrial applications. The results are shown in Table 4.3. The means are quite small for all T_1 to T_5, and the standard deviations are as low as 0.17, 1.59, 0.36, 0.38 and 0.68 for T_1 to T_5, respectively. This demonstrates the good performance of the developed modelling for the galvanizing process.

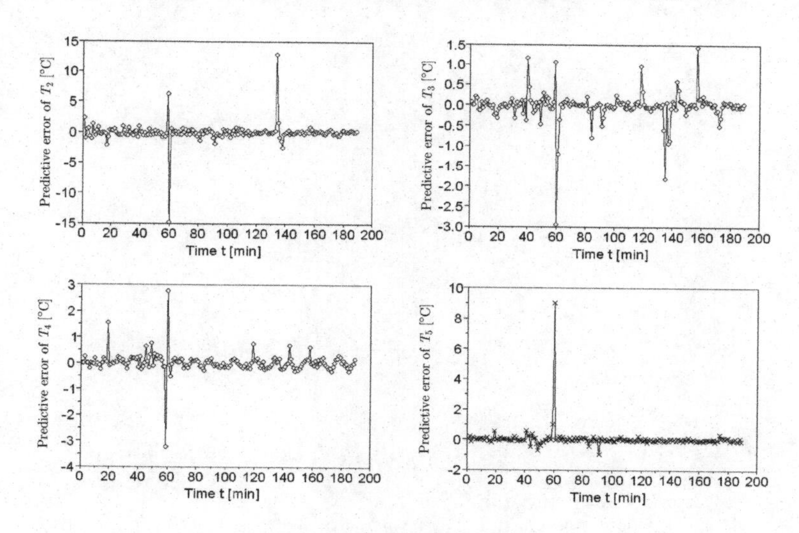

Fig. 4.10 Predictive errors of strip temperatures T_2, T_3, T_4 and T_5.

Table 4.3 Statistical analysis of the predictive errors of strip temperatures.

	T_1	T_2	T_3	T_4	T_5
Mean	0.0118519	0.0200000	-0.0305291	0.0228571	0.0466138
Std. dev.	0.1699820	1.5896832	0.3629016	0.3793027	0.6786627

Chapter 5

Wavelets-Based Methods

After discussing finite difference methods for solving ordinary differential equations (ODEs) and partial differential equations (PDEs) arising from modelling of complex industrial processes, this chapter is devoted to development of wavelet-based numerical methods for computation of process models, for which conventional numerical methods either fail or face difficulties.

Two types of wavelet-based numerical computation methods will be discussed: wavelet collocation method and wavelet Galerkin method. The wavelet collocation method is chosen because it has the simplest orthonormal series and also gives good numerical computation results. With useful mathematical properties, the wavelet Galerkin method has recently emerged as an accurate and efficient means of approximating the solution of PDEs. Application of wavelet Galerkin method over finite difference or element method have led to tremendous theoretical research and practical applications. The fundamental theories for both of these two wavelet-based methods have been been discussed previously in Chapter 2.

This chapter will apply wavelet collocation method in solving PDE models for complex chemical reaction processes, which are widely used in process industries. Model development, wavelet collocation method for model approximations, and numerical computation of the process model will be discussed.

We will also employ wavelet Galerkin method to numerically solving population balance equations arising from industrial crystallization processes, which are significant in fine chemical production, mineral processing and many other systems. Again, process modelling, wavelet Galerkin method for model approximations, and practical computation of the process model will be studied.

The materials presented in this chapter are partially taken from our previous publications [Yao *et al.* (1998, 2010a,b); Zhang *et al.* (2007)] and related preliminary studies.

5.1 Process Modelling for Chemical Reactions

In this section, chemical reaction processes, which are widely used in process industries, are considered as an example for development of wavelet collocation method. We will start with a brief introduction of the chemical reactions, particularly noncatalytic gas-solid reactions. This is followed by process modelling, leading to a set of coupled PDEs.

5.1.1 *Chemical Reactions*

Chemical reactions are processes that lead to transformation of one set of chemical substances, which are called reactants, into another, which involves products and/or other chemical substances. They are usually characterized by a chemical change, which yields one more products.

Noncatalytic gas-solid reactions occur in many important chemical and metallurgical processes as described in [Duduković and Lamba (1978); Ebrahimi and Jamshidi (2001); Ebrahimi *et al.* (2008)] and the references therein. Among many such chemical reactions are the calcination of limestone, production of iron in a blast furnace, the reduction of metallic oxides such as zinc oxide by methane, roasting of metallic sulfides, adsorption of acid gases by solid, activated carbon production, and coal gasification, to just name a few.

In order to better operate and control reaction processes, understanding the reaction mechanisms of the processes becomes crucial. Therefore, it is important to investigate the contributions of physical diffusion and chemical reaction towards the overall mass transfer rate, particularly in the case of non steady state reaction occurring in fixed bed equipment. Mathematical modeling of chemical reactions is a great help towards this end.

However, modelling of gas-solid reactions leads to a set of complex PDEs. Due to the complexity of the PDE model, analytical solutions are generally not available, implying that the PDE model system needs to be solved numerically. As conventional numerical methods fail to complete the numerical computation effectively and efficiently, this section introduces wavelet-based methods to deal with complex computing task.

5.1.2 *Model Development*

To show how the wavelet-based methods can be used for numerically solving the gas-solid reaction model, we consider a model in which a gas reacts with a solid to produce a gas product and solid product.

The scenario is as follows: Gas A diffuses and reacts with porous solid B to produce gas C and solid product D. The chemical reaction is is illustrated as follows

$$A + B \rightarrow C + D$$

During the reaction, effective gaseous diffusivity and oversize of the pellet do not change. The reaction process is isothermal and has reached pseudo-steady-state.

For model development, we use the following symbols and notations:

C_A: concentration of gas, A in the pellet;
C_{Ag}: concentration of gas, A in the bulk;
C_{B0}: initial solid concentration;
D_e: effective gas diffusivity in the porous solid;
F_p: shape factor of the pellet;
k: reaction rate constant;
r: position in the pellet;
R: characteristic pellet length;
t: time;
ν_B: stoichiometric coefficient of solid reactant.

Depending on the operating conditions and how the reaction mechanism is described, various mathematical models can be developed, which are generally formulated by coupled PDEs.

If we use a volume reaction model, then a dimensionless process model can be derived from the first principles [Duduković and Lamba (1978); Ebrahimi *et al.* (2008)] as follows

$$\frac{1}{x^{F_p}}\frac{\partial}{\partial x}\left(x^{F_p}\frac{\partial y}{\partial x}\right) - \phi^2 y F(z) = 0, \tag{5.1a}$$

$$\frac{\partial z}{\partial \theta} = -yF(z), \tag{5.1b}$$

where θ, x, y and z are the dimensionless time, position in the pellet, gas concentration and solid concentration, respectively, and are defined as

$$x = \frac{r}{R}, \tag{5.2}$$

$$y = \frac{C_A}{C_{Ag}}, \tag{5.3}$$

$$\theta = k\nu_B \frac{C_{Ag} f(C_{B0}) t}{C_{B0}}. \tag{5.4}$$

In addition, there are also two functions, ϕ and F, in Equation (5.1). The function ϕ is the so-called Thiele modulus, which is defined by

$$\phi = R\sqrt{k\frac{f(C_{B0})}{De}}. \tag{5.5}$$

F is a function on solid concentration, and is given as

$$F(z) = z^p,\ p \geq 0 \tag{5.6}$$

The initial and boundary conditions of the PDE model in Equation (5.1) are given as follows:

$$\begin{cases} \theta = 0: & z = 1; \\ x = 0: & \dfrac{\partial y}{\partial x} = 0; \\ x = 1: & \dfrac{\partial y}{\partial x} = Bi_m(1 - y). \end{cases} \tag{5.7}$$

Due to the complexity from the PDE coupling, the gas-solid model described by Equations (5.1) and (5.7) is hard to solve. In order to overcome this difficulty, Duduković and Lamba (1978) introduced the concept of **Cumulative Gas Concentration**, which is defined by

$$Y(x, \theta) = \int_0^\theta y(x, \theta) d\theta. \tag{5.8}$$

Combining Equation (5.8) with Equation (5.1b) gives

$$z = \psi^{-1}(-Y). \tag{5.9}$$

By introducing these new variables, the original model Equations (5.1) and (5.7) can be reduced to the following form

$$\frac{1}{x^{F_p}} \frac{\partial}{\partial x}\left(x^{F_p} \frac{\partial Y}{\partial x}\right) - \phi^2[1 - \psi^{-1}(-Y)] = 0, \tag{5.10}$$

with the boundary conditions

$$\begin{cases} x = 0: & \dfrac{\partial Y}{\partial x} = 0; \\ x = 1: & \dfrac{\partial Y}{\partial x} = Bi_m(\theta - Y). \end{cases} \tag{5.11}$$

Equations (5.10) and (5.11) are our final mathematical model of the chemical reaction process. Next, we will numerically solve this model by using wavelet collocation method.

5.2 Three Versions of Wavelet Collocation Methods

Since the introduction of the Haar wavelet function, many forms of wavelet functions have been developed, including Shannon, Daubechies and Legendre wavelets. Among those forms, Haar wavelets have the simplest orthonormal series with compact support. This makes Haar wavelets particularly attractive, motivating the application of wavelet collocation method in solving complex industrial process models.

Before applying the wavelet collocation technique to the reaction process models, as done in [Bertoluzza and Naldi (1996)], this section briefly introduces three versions of wavelet collocation methods.

5.2.1 *Basic Wavelet Collocation Method*

As indicated in Section 2.7, interpolating wavelets should be used as the basis functions when the collocation method is our option for solving differential equation models.

From section 2.7, the interpolating wavelets are denoted by $\theta_{j,k}$, and then any smooth function $f \in V_J$ can be approximated as the expansion of $\theta_{j,k}$:

$$f(x) \doteq \sum_{k=-\infty}^{\infty} f_{J,k}\theta_{J,k} \tag{5.12}$$

with $f_{J,k}$ is the value of f at the kth dyadic point.

5.2.2 *Improved Wavelet Collocation Method I*

In order to improve the accuracy of the approximation and/or deal with boundary conditions, Equation (5.12) can be further modified. Precisely, we introduce modified wavelet basis based on the interpolating wavelets discussed in Section 2.7 as follows:

$$\begin{cases} \widetilde{\theta}_{j,0} = \sum_{k=-\infty}^{0} \theta_{j,k}, \\ \widetilde{\theta}_{j,k} = \theta_{j,k} \quad \text{for} \quad k = 1, \cdots, 2^j - 1, \\ \widetilde{\theta}_{j,2^j} = \sum_{k=2^j}^{+\infty} \theta_{j,k}. \end{cases} \tag{5.13}$$

Obviously, the functions $\widetilde{\theta}_{j,k}, k = 0, \cdots, 2^j$, have similar properties to those of $\theta j, k$. Therefore, $\{\widetilde{\theta}_{j,k}\}$ can be used as basis functions.

Thus, Equation (5.12) has the form of

$$f(x) \doteq \sum_{k=0}^{2^J} f_{J,k}\theta_{J,k}. \tag{5.14}$$

It is noticed that there are $2^J + 1$ unknowns to be determined if Equation (5.14) is used to solve the differential equations. With consideration of the compact support of the basis function $\theta(x)$, Equation (5.14) can have the form of

$$f(x) \doteq \sum_{k=-L+1}^{2^J+L-1} f_{J,k}\theta_{J,k}, \tag{5.15}$$

in which $2^J + 2L - 1$ coefficients need to be determined.

However, we have only collocation points at the $2^J + 1$ dyadic points. This means that we need $2L - 2$ more points for determination of the values of these coefficients. Considering the boundary conditions, we add $L - 1$ more points near each boundary, and denote them by $x_k, k = -L + 1, \cdots, -1$ and $x_k, k = 2^J + 1, \cdots, 2^J + L - 1$, respectively. This makes the total number of points the same as the number of unknowns to be determined, implying that the problem becomes solvable.

5.2.3 *Improved Wavelet Collocation Method II*

Another improved wavelet collocation scheme is based on the improved method I proposed in the last subsection. In order to improve the way dealing with the boundary conditions, instead of using the values of f at points near the boundaries $x_k, k = -L + 1, \cdots, -1, k = 2^J + 1, \cdots, 2^J + L - 1$, using the values of f which are extrapolated from these points. In order to do so, we define

$$a_k(x) = \prod_{\substack{i=0 \\ i \neq k}}^{2M-2} \frac{x - x_i}{x_k - x_i}, \quad b_k(x) = \prod_{\substack{i=2^J-2M+2 \\ i \neq k}}^{2^J} \frac{x - x_i}{x_k - x_i}. \tag{5.16}$$

It is easy to verify that

$$a_{nk} = a_k(x_n) = \begin{cases} 1 & \text{if } n = k, \\ 0 & \text{otherwise} \end{cases} \tag{5.17}$$

$$b_{nk} = b_k(x_n) = \begin{cases} 1 & \text{if } n = k, \\ 0 & \text{otherwise} \end{cases} \tag{5.18}$$

Furthermore, we define two Lagrange polynomial functions P_1 and P_2 with order of $2M-1$ to satisfy

$$P_1(x_n) = \sum_k a_{nk} f(x_k),$$
$$n = -L, \cdots, -1; k = 0, \cdots, 2M-1 \quad (5.19)$$
$$P_2(x_n) = \sum_k b_{nk} f(x_k),$$
$$n = 2^J+1, 2^J+L; k = 2^J-2M+2, \cdots, 2^J \quad (5.20)$$
$$P_1(x_k) = f(x_k), k = 0, \cdots, 2M-1 \quad (5.21)$$
$$P_2(x_k) = f(x_k), k = 2^J-2M+1, \cdots, 2^J \quad (5.22)$$

With these two defined polynomials, Equation (5.12) reads

$$f(x) \doteq \sum_{n=-L}^{-1} P_1(x_n)\theta_{J,n} + \sum_{k=0}^{2^J} f(x_k)\theta_{J,k} + \sum_{n=2^J+1}^{2^J+L} P_2(x_n)\theta_{J,n}. \quad (5.23)$$

Rearranging Equation (5.23) and introducing

$$\theta^l_{jk} = \theta_{jk} + \sum_{n=-L}^{-1} a_{nk}\theta_{jn}$$
$$\theta^r_{jk} = \theta_{jk} + \sum_{n=2^J+1}^{2^J+L} b_{nk}\theta_{jn}$$

yield the improved wavelet collocation method II for approximating function $f(x)$:

$$f(x) \doteq \sum_{k=0}^{L} f(x_k)\theta^l_{J,k} + \sum_{k=L+1}^{2^J-L-1} f(x_k)\theta_{J,k} + \sum_{k=2^J-L}^{2^J} f(x_k)\theta^r_{J,k}. \quad (5.24)$$

5.3 Wavelet Collocation Method for Reaction Processes

This section demonstrates how to use the wavelet-based collocation method to numerically solve PDEs (5.10) and (5.11) by using improved wavelet method I.

To avoid any confusion, we denote the dimensionless time θ by t. The unknown function Y can then be approximated by

$$Y(x,t) = \sum_{k=0}^{2^j} Y_{j,k}(t)\widetilde{\theta}_{j,k}(x). \quad (5.25)$$

The boundary conditions are reduced to

$$\sum_{k=0}^{2^j} Y_{j,k}(t)\widetilde{\theta}_{j,k}^{(1)}(x) = 0$$
$$Y_{j,2^j} = t \tag{5.26}$$

From Equations (5.1b), (5.7) and (5.8), we have

$$z^{1-p} = 1 - (1-p)Y \quad \text{for} \quad q \neq 1 \tag{5.27}$$

Consider three rate forms: a first-order dependence on the gas reactant concentration and zeroth ($p = 0$), a half dependance on solid reactant concentration ($p = 1/2$), and a first-order dependence on solid reactant concentration ($p = 1$). From Equations (5.1b) and (5.27), we have

$$z = 1 - Y \text{ for } p = 0, \tag{5.28a}$$

$$z = \left(1 - \frac{1}{2}Y\right)^2 \text{ for } p = 1/2, \quad \text{and} \tag{5.28b}$$

$$z = e^{-Y}, \text{ for } p = 1, \tag{5.28c}$$

respectively.

If the zero order volume reaction model with respect to solid reactant is considered, the model Equation (5.10) becomes

$$\frac{\partial^2 Y}{\partial x^2} + \frac{F_p}{x}\frac{\partial Y}{\partial x} - \phi^2 Y = 0. \tag{5.29}$$

The solid conversion, η, is a very important parameter in engineering applications. It can be calculated from the following equation

$$\eta = 1 - (F_p + 1)\int_0^1 x^{F_p} z dx. \tag{5.30}$$

As we know, shape factor F_p has the values of 0, 1 and 2, which correspond to slab, cylinder and sphere, respectively. Considering Equation (5.26), for shape factor $F_p = 0$, we can calculate the solid conversion η by the following relationship

$$\eta = 1 - \int_0^1 z dx = \sum_{k=0}^{2^j} Y_{j,k}\int_0^1 \widetilde{\theta}_{j,k} dx$$
$$= Y_{j,0}\sum_{k=-L+1}^{0}\int_0^1 \theta_{j,k} dx + \sum_{k=1}^{2^j-1} Y_{j,k}\int_0^1 \theta_{j,k} dx + Y_{j,2^j}\sum_{k=2^j}^{2^j+L-1}\int_0^1 \theta_{j,k} dx \tag{5.31}$$

with

$$\int_0^1 \theta_{j,k} dx = \frac{1}{2^j} \left(I(2^j - k) - I(-k) \right) \tag{5.32}$$

and $I(x)$ defined by

$$I(x) = \int_{-\infty}^{x} \theta(y) dy \tag{5.33}$$

For the chemical reaction process model under investigation, simulations have been conducted for numerical computation of the model by using the numerical solutions based on the wavelet collocation technique developed in this section. For Thiele modulus $\phi = 1, 5, 10, 25$, the numerical solutions are depicted in Figure 5.1.

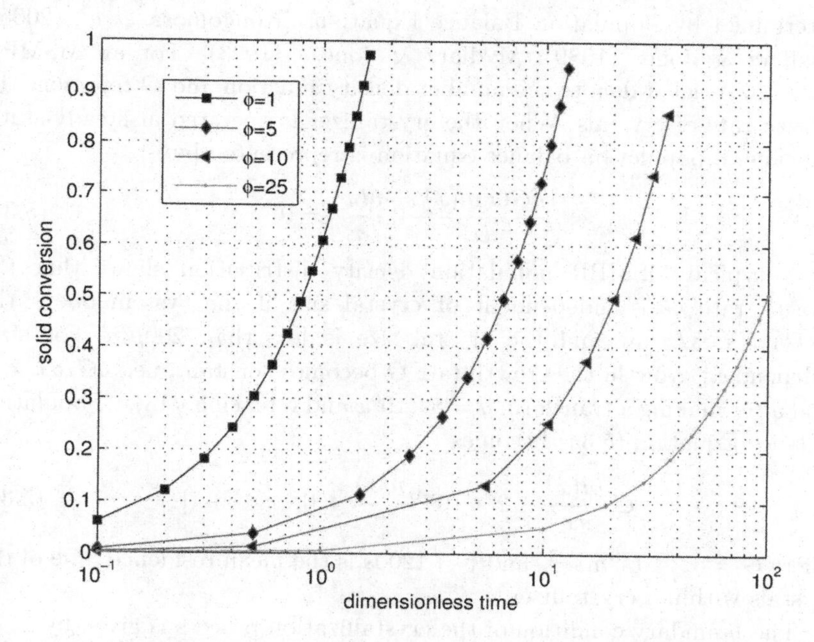

Fig. 5.1 Zero-order volume reaction model with slab pellet for different Thiele modulus.

5.4 Model Development for Crystallization Processes

This section aims to establish mathematical models for industrial crystallization processes. This will be followed in the next few sections by investigations into wavelet Galerkin method for numerical computation of the crystallization process models.

Crystallization is the process of formation of crystals precipitating from a solution or melt. It also refers to a chemical solid-liquid separation technique widely used in various industrial systems. It appears in production of fine crystals, mineral processing, food production, and many other industrial systems.

Main crystallization processes employed in industries include cooling crystallization and evaporative crystallization. We consider a Mixed Suspension Mixed Product Removal (MSMPR) cooling crystallizer as an example, which consists of well-mixed, continuously operated vessels.

The crystallization kinetics of an MSMPR cooling crystallizer is usually determined by Population Balance Equations [Kougoulosa *et al.* (2005); Mydlarz & Jones (1989); Mydlarz & Jones (1993)]. For an MSMPR crystallizer, let n denote the number density function and G represent the growth rate of crystals. When the crystallizer is operated at steady-state, the general population balance equation is represented by

$$\frac{dG(x)n(x)}{dx} + \frac{n(x)}{\tau} = 0. \tag{5.34}$$

A typical MSMPR population density distribution shows that the growth rate G is independent of crystal size if the system obeys the McCabe's ΔL law and the crystal size is less than $200\mu m$. For size-independent growth, the growth rate G becomes constant, i.e., $dG/dx = 0$. And after making a transform $y = 5\times 10^5 x$ and replacing y by x, Population Balance Equation (5.34) becomes

$$G\frac{dn(x)}{dx} + 2\times 10^{-6}\frac{n(x)}{\tau} = 0, \;\; x \in [0,1] \tag{5.35}$$

where $G = 5\times 10^{-8} ms^{-1}$, and $\tau = 1200s$ is the mean residence time of the crystals within a crystallizer.

The boundary condition of the crystallization process is given by

$$\ln(n^0) = 34, \quad \text{nuclei population density} \tag{5.36}$$

Therefore, we have derived an ODE model in Equation (5.35) with the boundary condition in Equation (5.36) for an MSMPR crystallizer. It is worth mentioning that for more complex crystallization processes, PDEs will have to be used to describe the crystallization kinetics. The wavelet Galerkin method to be investigated below is suitable for general PDE models, and thus can handle general crystallization process models.

5.5 Wavelet Galerkin Method for PDEs

Consider a PDE for function $u(x_1, \cdots, x_n)$ in the following form

$$F(x_1, \cdots, x_n, u, \frac{\partial u}{\partial x_1}, \cdots, \frac{\partial u}{\partial x_n}, \frac{\partial^2 u}{\partial x_1^2}, \frac{\partial^2 u}{\partial x_1 \partial x_2}, \cdots) = 0. \tag{5.37}$$

The wavelet scaling function expansion up to order of j reads

$$\begin{aligned} u(t,x) &= \sum_{k=-\infty}^{\infty} u_{j,k}(t)\phi_{j,k}(x) \\ &= \sum_{k=-\infty}^{\infty} u_{j,k}(t)\phi(2^j x - k). \end{aligned} \tag{5.38}$$

If the scaling function is the one for Daubechies orthonormal wavelet with compact support $[0, L]$, Equation (5.38) can be approximated as

$$u(t,x) = \sum_{k=2-L}^{2^j} u_{j,k}(t)\phi_{j,k}(x). \tag{5.39}$$

This is a projection of the solution of (5.37) onto subspace V_j with basis $\phi_{j,k}(x)$.

The basis function $\phi_{j,k}$ can be replaced with the basis functions of W_j, $\psi_{j,k}$, or the basis functions of $V_j \oplus W_{j+1}$ $\{\phi_{j,k}, \psi_{j+1,k}\}$. In the Galerkin method, using a pure wavelet basis, a wavelet-scaling function basis, and a pure scaling function basis expansion are completely equivalent. Each basis defines the same space of approximation. The Galerkin coefficients associated with each basis are related by orthogonal transformations determined by a unitary change of basis. However, it is mathematically simpler to use the pure scaling function expansion.

To uniquely determine the coefficients appeared in Equation (5.39), we substitute Equation (5.39) into Equation (5.37) and again project the resulting expressions, each of which is orthogonal to others, onto the subspace V_j. It follows that

$$\int_{-\infty}^{\infty} F(x_1, \cdots, x_n, u, \frac{\partial u}{\partial x_1}, \cdots, \frac{\partial u}{\partial x_n}, \frac{\partial^2 u}{\partial x_1^2}, \frac{\partial^2 u}{\partial x_1 \partial x_2}, \cdots)\phi_{jk}(x)dx = 0. \tag{5.40}$$

This equation can be evaluated based on the connection coefficients defined in Section 2.7.3.

5.6 Solution Based on Wavelet Galerkin Method

To demonstrate how the wavelet Galerkin method works, we consider an example of the population balance model in Equation (5.34) or (5.35) with the boundary condition in Equation (5.36). The model is developed for a crystallizer. The wavelet Galerkin method will be used to numerically solving the process model.

Let us consider model Equation (5.34). By the wavelet Galerkin scheme, the number density function $n(t, x)$ can be approximated as the expansion of scaling function and a finite number of its translates. We have

$$n(x) = \sum_k n_k \phi_{j,k}(x) \tag{5.41}$$

where n_k for different values of k are coefficients to be determined. It follows from Equation (5.40) and the boundary condition in Equation (5.36) that

$$G \sum_{k=-L+2}^{2^j-1} n_k C^1_{j,l,k} + \frac{2 \times 10^{-4}}{\tau} \sum_{k=-L+2}^{2^j-1} n_k C^0_{j,l,k} = 0$$
$$\text{for} \quad l = 2 - L, \cdots, 2^j - 1 \tag{5.42}$$

and

$$\sum_{k=-L+2}^{2^j-1} n_k \phi(-k) = n^0, \tag{5.43}$$

where the connection coefficients have been defined as in Section 2.7.3 and have the following expressions

$$C^1_{j,l,k} = \int_0^1 \phi_{j,l}(x) \frac{d\phi_{j,k}(x)}{dx} dx, \tag{5.44}$$

$$C^0_{j,l,k} = \int_0^1 \phi_{j,l}(x) \phi_{j,k}(x) dx. \tag{5.45}$$

Then, for $l = 2 - L, \cdots, 2^j - 1$, Equations (5.42) through (5.45) give

$$\sum_{k=-L+2}^{2^j-1} n_k \left(G C^1_{j,l,k} + \frac{2 \times 10^{-4}}{\tau} C^0_{j,l,k} + \phi(-k) \right) = n^0. \tag{5.46}$$

Together with the algorithms for computing the connection coefficients in Section 2.7.3, Equation (5.46) gives a numerical solution to the population balance model in Equation (5.34).

Simulation results of the wavelet Galerkin method for the population balance equation model are depicted in Figure 5.2. It is seen from Figure 5.2

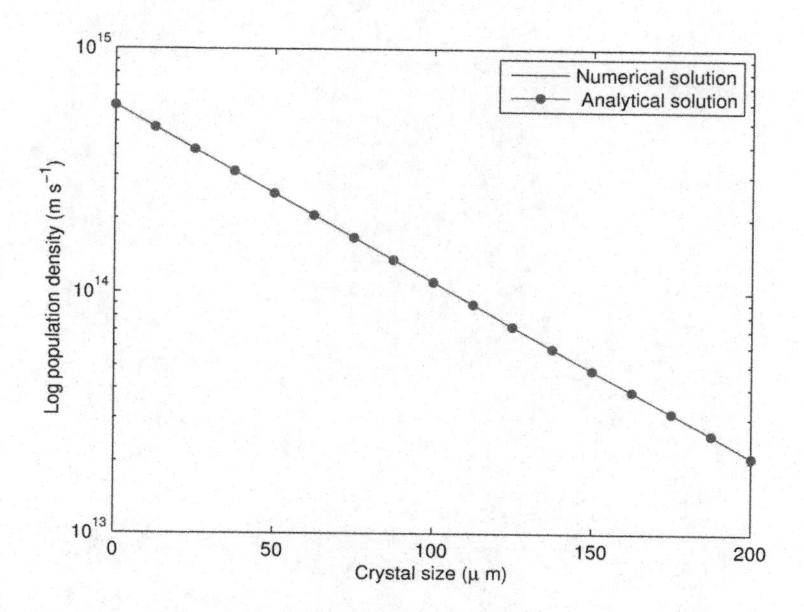

Fig. 5.2 Results of the MSMPR cooling crystallizer model holding McCabe's ΔL law.

that the numerical computation results from the wavelet Galerkin method match the analytical results very well without visible difference on the plot.

In this section, the proposed wavelet Galerkin method has been used to study the numerical solution to a crystalisation process model governing by an ODE. In this case, substituting the approximation (5.41) into the governing Equation (5.35) results in a set of algebraic equations expressed in Equation (5.42).

However, if the governing equation is a PDE, the substitution will lead to a set of ODEs with a set of algebraic equations from the boundary condition. Then, the system of the ODEs can be numerically solved by using numerical computation techniques developed for ODEs. For example, the wavelet Galerkin method discussed in this section, or the numerical methods introduced in Chapter 2 (e.g., the Runge-Kutta method in Section 2.5), can be considered as a numerical computation tool.

Chapter 6

High Resolution Methods

This chapter develops high resolution methods for numerical computation of mathematical models for complex industrial processes. We consider two typical types of complex industrial processes as examples to show how high resolution methods work for numerically solving process models: column chromatographic separation and population balance equation problems. For each of these two types of processes, we will discuss process modelling, model approximations and numerical computation of the process model.

Materials presented in this chapter are partially taken from our publications [Zhang *et al.* (2008); Yao *et al.* (2010b)].

6.1 Column Chromatographic Separation Processes

Chromatography refers to a set of laboratory techniques for the separation of mixtures. The mixture dissolved in a fluid known as mobile phase; while the substance fixed in place for the chromatography procedure is called stationary phase. When various constitutes of the mixture moves at different speeds, the separation of the constitutes occurs because of the differential partitioning between the mobile phase and stationary phase.

Column chromatography is a chromatographic separation technique in which the stationary bed is within a tube. Similar to general chromatographic techniques, column chromatography is also typically used in laboratories for efficient separation of mixtures of chemical compounds. The basic principle of the separation is the use of the phenomenon of adsorption, i.e., the components in the milieu of solvent and adsorbent exhibit different adsorption behaviours.

The specific separation method of "Column Chromatography" typically uses a glass column filled with adsorbent through which a composite liquid mixture passes. The technique operates such that each component will form in a different section of the column arranged by colour according to the adsorption affinity of each material. If an appropriate desorbent is poured into the column, the components that have been adsorbed on the adsorbent dissolve into the desorbent and start moving downwards in the column with different migration rates. The components in the lower layers move faster. The liquids can then be collected from the bottom of the column as they drip out. This mechanism is illustrated in Figure 6.1.

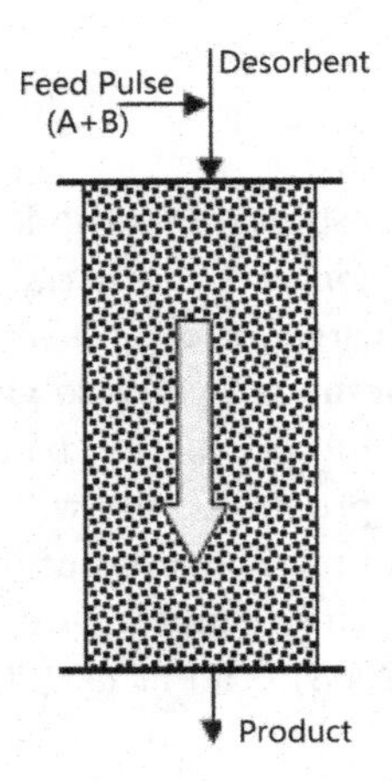

Fig. 6.1 Column chromatographic separation.

Chromatography was initially developed for extraction and purification of complex mixtures of vegetal origin; and later it achieved rapid growth and has now become a ubiquitous analytical method. The strengths of the technique are especially manifested in the separation of isomers and natural materials.

A single chromatographic column model describes the kinetics of adsorption. It is the combination of generally formatted equations with any possible isotherm expression. The level of mathematical difficulty encountered depends much on the nature of equilibrium relationship, the concentration level, and the choice of flow mode. Mathematical models of varying complexity have been used to investigate the separation performance of the chromatography. The information of the exact position and shape of the transient concentration fronts as a function of fluid velocity and component concentrations is vital for design and optimization of a chromatographic process.

6.2 Model Development for Column Chromatography

6.2.1 *Mechanism and Assumptions for Process Modelling*

In a separation column, the mobile phase containing the solvent and components flows through the solid phase, the package of fine, globular and porous particles (adsorbent). As shown in Figure 6.2, the particle consists of a solid part and a liquid pore phase, and is surrounded by a film. The affinity of the component to the adsorbent leads to the transport of the component from the solvent to the particles. The driving force of this transport is the tendency of equilibrating the molecular load between the solvent and the surface of the particles.

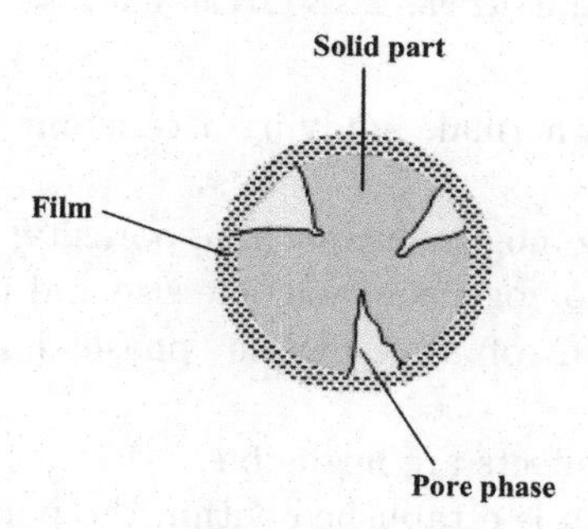

Fig. 6.2 Adsorbent particle.

The mass transfer between the mobile and stationary phases is dominated by three phenomena: convection, diffusion and dispersion. Convective and dispersive mass transfers are present in the mobile phase. The mass exchange between the mobile and solid phase takes place by diffusion through the film of solvent surrounding the particle. In principle, adsorptive exchange takes place all over the particle surface. However, for ease of modelling, only the pore phase is supposed to be in exchange with the particle surface, and the exchange between the pore phase and the mobile phase occur through the film.

The mechanism of mass transfer and mass exchanges in a chromatographic separation process is illustrated in Figure 6.3.

Before we derive mathematical descriptions for the adsorptive chromatographic separation process, several assumptions are made as follows, which are common in theoretic research of column chromatography:

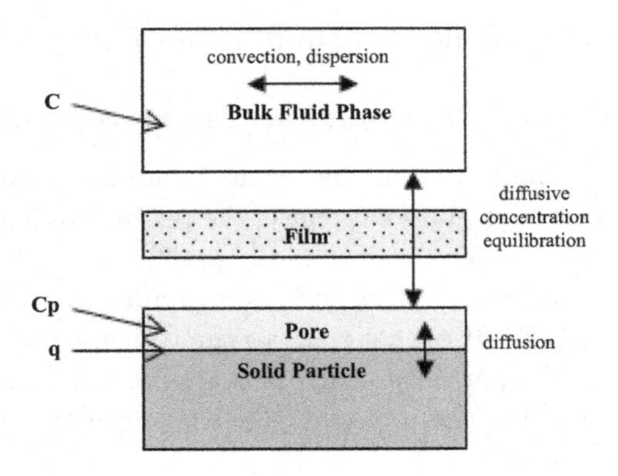

Fig. 6.3 Mass transfer and mass exchange in a separation process.

- The process is isothermal, implying a constant temperature during a run;
- The column has homogeneous package porosity;
- The column has homogeneous particle size and porosity;
- The concentration of the mobile phase has a constant radial distribution;
- The column wall effects are negligible;
- A local equilibrium is established within the pores;
- The film mass transfer can be used to describe the interfacial mass transfer between the bulk-fluid and particle phases; and
- The diffusion and mass transfer parameters are constant.

With the above assumptions and from fluid dynamics and mass transfer, mathematical descriptions can be established for the three phenomena described previously, i.e., convection, diffusion and dispersion for the mass transfer between the mobile and stationary phases.

6.2.2 *Basic Model*

Among different types of mathematical models for column chromatography, the Transport-Dispersive Linear Driving Force (LDF) modelling framework has been widely used in modelling many adsorption processes due to their simplicity and good agreement with experimental results. A Transport-Dispersive model takes into account the major mass transfer kinetics arising in chromatography by including the axial dispersion, mass transfer resistance and the diffusion between the mobile and stationary phases. The mass balance is written for a fluid percolating through a bed of spherical particles in the column. Furthermore, the Transport-Dispersive modelling assumes that the mass transfer kinetics between the mobile phase and pore phase percolating across the column and particles is infinitely fast. In addition, local mass transfer resistance can be neglected.

Therefore, we have the following model equation

$$\frac{\partial C}{\partial t} + u\frac{\partial C}{\partial x} = D_r\frac{\partial^2 C}{\partial x^2} - \frac{1-\epsilon_T}{\epsilon_T}\frac{\partial q}{\partial t}, \tag{6.1}$$

where C and q are concentrations of liquid phase and solid phase, respectively; u stands for interstitial velocity, D_r is a lumped dispersion and diffusion coefficient, and ϵ_T represents total porosity.

The last term of the right-hand side of Equation (6.1) involves partial derivative $\frac{\partial q}{\partial t}$ for the solid phase concentration. This partial derivative reflects the kinetic effects of mass transfer on the system behaviour. It is commonly described by a Linear Driving Force (LDF) Model as

$$\frac{\partial q}{\partial t} = k_{\text{eff}}(q^* - q), \tag{6.2}$$

where k_{eff} stands for effective mass transfer resistance, and q^* is the equilibrium concentration in the interface between the two phases.

The LDF model in Equation (6.2) is an application of the first Fick's law of mass transport, and is an equation of equilibrium isotherm. Offering a realistic representation of industrial processes, it is shown to be a good compromise between accurate and efficient solutions of these models [Biegler *et al.* (2004)].

Equations (6.1) and (6.2) form the basic mathematical model of the column chromatographic process.

6.2.3 *Adsorption Isotherm Model*

In the basic mathematical model shown in Equations (6.1) and (6.2), the equilibrium concentration $q*$ in the interface between liquid and solid phases is not a constant. It changes over time when the operating condition of the chromatographic process changes. An adsorption isotherm model is developed to describe the dynamics of $q*$.

An isotherm model relates the component concentration in solid phase to the one in mobile phase. It describes the thermodynamic behaviour of the modelled system. Influencing the structure of the resulting mathematical expressions, the type of adsorption isotherms is a key to distinguish different chromatographic processes. Process with linear isotherms are most applied for sugar separation. It can be modelled and simulated much easier than those with nonlinear isotherms, e.g., enantiomers separation.

The adsorption equilibrium isotherm can be expressed in a general form:

$$q_i^* = f_i(C_1, C_2, \cdots, C_M), \tag{6.3}$$

where M stands for the total number of components in feeding mixture. The following are some of the equilibrium isotherm models reported in the literature.

Freundlich:

$$q_i^* = kC_i^{1/n}; \tag{6.4a}$$

Langmuir:

$$q_i^* = \frac{bC_i}{1 + bC_i}; \tag{6.4b}$$

Extended Langmuir:

$$q_i^* = \frac{a_i C_i}{1 + \sum_{n=1}^{M} b_n C_n}; \tag{6.4c}$$

For ion-exchange:

$$q_i^* = \frac{a_i C_i}{\sum_{n=1}^{M} b_n C_n}; \tag{6.4d}$$

Langmuir-Freundlich:

$$q_i^* = \frac{a_i C_i^{m_i}}{1 + \sum_{n=1}^{M} b_n C_n^{m_n}}; \tag{6.4e}$$

Modified Competitive Langmuir:

$$q_i^* = \lambda_i C_i + \frac{a_i C_i}{1 + \sum_{n=1}^{M} b_n C_n}. \tag{6.4f}$$

6.2.4 *Complete Process Model*

With the basic model and adsorption isotherm model established above, a complete chromatographic column model is formed by Equations (6.1) and (6.2) plus the isotherm model in Equation (6.3). For a specific chromatographic column, the isotherm model can be chosen from Equation (6.4).

Analytical solutions to such a chromatographic process model are difficult to find. This chapter will use high resolution method to numerically solving the model equations. In order to evaluate the performance of the high resolution method, we consider a simple linear equilibrium case. For this case, an analytical solution is given by [Lapidus and Amundson (1952)] for any time at any point in the column.

In the case where equilibrium is established at each point in the bed, i.e.,

$$q = q^*, \tag{6.5}$$

Equation (6.1) can be transformed into

$$\frac{\partial C}{\partial t} = \frac{1}{1 + \dfrac{1}{\alpha}\dfrac{\partial q^*}{\partial C}}(-u\frac{\partial C}{\partial x} + D_r\frac{\partial^2 C}{\partial x^2}), \tag{6.6}$$

where α is the fractional void volume

$$\alpha = \frac{\epsilon_T}{1 - \epsilon_T}. \tag{6.7}$$

In addition, for a linear case, the equilibrium isotherm has the following simple expression

$$q^* = kC. \tag{6.8}$$

It follows that $\dfrac{\partial q^*}{\partial C} = k$. Substituting this relationship into Equation (6.6) gives a simple model equation for column chromatography as follows

$$\frac{\partial C}{\partial t} = \frac{1}{1 + \dfrac{k}{\alpha}}(-u\frac{\partial C}{\partial x} + D_r\frac{\partial^2 C}{\partial x^2}). \tag{6.9}$$

It is seen that the right hand side of Equation (6.9) includes the convective and dispersive terms, both of which are weighted by a factor involving the phase relation and the slope of the isotherm. This also implies that the effects of axial dispersion and mass transfer are additive in linear.

6.3 Analytical Solution for Linear Equilibrium Case

The mathematical model in Equation (6.9) for column chromatographic process in linear equilibrium case is solvable analytically. Let us develop an analytical solution to the process model.

Consider a pulse injection with a concentration of C_0 and duration of t_0, and specify two constant concentrations as boundary conditions:

$$C(0,t) = \begin{cases} C_0, \ t \leq t_0 \\ 0, \ t > t_0 \end{cases} \tag{6.10}$$

We also specify the initial conditions

$$C(x,0) = 0 \tag{6.11}$$

We also assume the chromatographic column has infinite extent, implying that the end boundary condition is not included. Then, we investigate the concentration profiles at column length of L.

Using dimensionless transformation, let

$$\tau = ut\frac{\alpha}{L}, Pe = u\frac{t}{D_r}, z = \frac{x}{L}, c = \frac{C}{C_0}, \widetilde{q} = \frac{q}{C_0}. \tag{6.12}$$

Then, Equation (6.9) becomes

$$\frac{\partial c}{\partial \tau} = \frac{1}{\alpha + k}(-\frac{\partial c}{\partial z} + \frac{1}{Pe}\frac{\partial^2 c}{\partial z^2}) \tag{6.13}$$

with boundary conditions

$$c(0,\tau) = \begin{cases} 1, \ \tau \leq \tau_0 \\ 0, \ \tau > \tau_0 \end{cases} \tag{6.14}$$

and initial conditions

$$c(z,0) = 0. \tag{6.15}$$

From reference [Lapidus and Amundson (1952)], the analytical solution to the simplified model in Equations (6.13), (6.14) and (6.15) in linear equilibrium case is

$$c(z,\tau) = \begin{cases} H(\tau), \ \tau \leq \tau_0 \\ H(\tau) - H(\tau - \tau_0), \ \tau > \tau_0 \end{cases} \tag{6.16}$$

with

$$H(\tau) = \frac{1}{2}\left[1 + \mathrm{erf}\left(\sqrt{\frac{\tau Pe}{4(\alpha+k)}} - z\sqrt{\frac{Pe(\alpha+k)}{4\tau}}\right)\right.$$
$$\left. + e^{zPe}\mathrm{erfc}\left(\sqrt{\frac{\tau Pe}{4(\alpha+k)}} + x\sqrt{\frac{Pe(\alpha+k)}{4\tau}}\right)\right] \tag{6.17}$$

where erf() and erfc() are the error and complementary error functions, respectively,

$$\text{erf}(x) = \frac{2}{\sqrt{\pi}} \int_0^x e^{-t^2} dt, \tag{6.18}$$

$$\text{erfc}(x) = \frac{2}{\sqrt{\pi}} \int_x^\infty e^{-t^2} dt. \tag{6.19}$$

The analytical solution from Equation (6.16) under given parameters is graphically depicted in Figure 6.4. Figure 6.4 shows the dynamics of the adsorption process of the column chromatography. Specifically, it shows that the adsorbate is gradually adsorbed by the adsorbent and its concentration in liquid is decreasing along the column length.

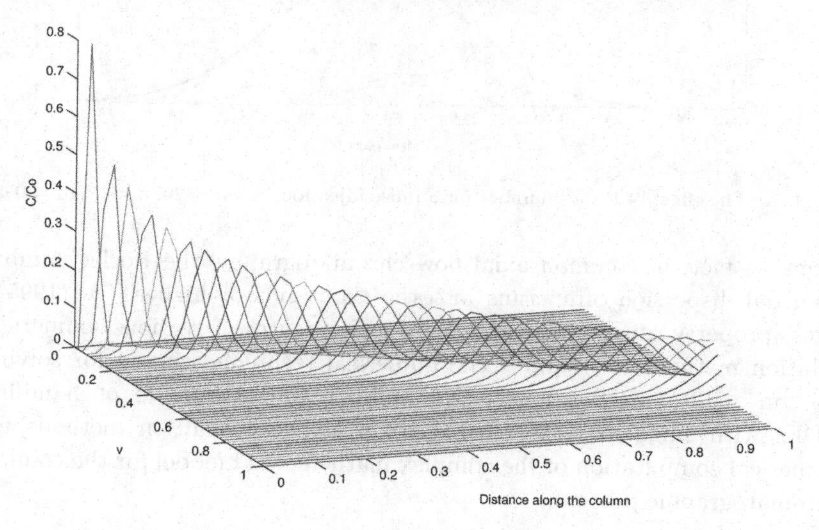

Fig. 6.4 Analytical solution: $Pe = 500, \alpha + k = 20$.

One of the factors that determine the property of the concentration wave is the Péclet number Pe. In fluid dynamics, the Péclet number is a dimensionless number relating the rate of advection of a flow to its rate of diffusion. Figure 6.5 shows the liquid concentration distributions at a certain time instant under different Péclet numbers. Other parameters are kept unchanged in plotting this figure. It is seen from Figure 6.5 that with the increasing of Pe, the shape of the wave gets steeper. In another words, large Pe values contribute to stiff concentration profiles.

For theoretical studies, it is not necessary to run very stiff cases. However, in engineering applications, the Péclet number is often very

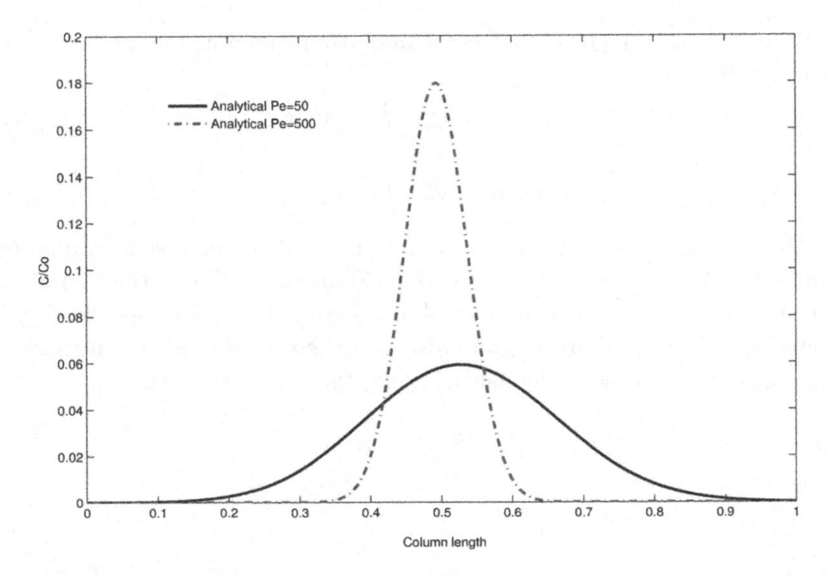

Fig. 6.5 The effect of Péclet number for a pulse injection ($\alpha + k = 20, \tau = 10, \tau_0 = 0.4$).

large. In fact, in common axial flow chromatography, the Péclet number for axial dispersion often runs into the thousands or higher [Gu (1995)]. This property directly affects the performance of various numerical solution methods. It requires the numerical technique chosen for solving column model with a higher Pe number to be capable of handling PDEs with singularity. We will develop a high resolution methods for numerical computation of the complex mathematical model for the column chromatographic process.

6.4 Model Discretization Using High Resolution Methods

In wavelet collocation method that we discussed in Chapters 2 and 5, the concerned interval is discretized by using the dyadic points. The discretization can take advantage of the interpolation property of the interpolating wavelet. In comparison, high resolution numerical schemes use flux/slope limiters to limit the gradient around shocks or discontinuities.

High Resolution schemes are used in the numerical solution of PDEs where high accuracy is required in the presence of shocks or discontinuities. They can achieve second- or higher-order spatial accuracy with fewer mesh points in comparison with other first-order numerical computing schemes with similar accuracy. With a high resolution scheme, solutions to PDEs are free from spurious oscillations, and high accuracy can be obtained around the shocks.

In the following, we will use the high resolution method to find the numerical solution to the PDE model represented by Equations (6.13), (6.14) and (6.15). Three cases will be studied corresponding to $Pe = 50, 500$ and 5000, respectively.

Consider PDE (6.13) with appropriate boundary and initial conditions. Divide the concerned interval $z \in [0, 1]$ into a number of subintervals uniformly. For N subintervals $\Omega_1, \Omega_2, \cdots, \Omega_N$, let

$$z = 1/N, z_{1/2} = 0, z_{N+1/2} = 1. \tag{6.20}$$

We have

$$\Omega_i = [z_{i-1/2}, z_{i+1/2}], \ i = 1, \cdots, N \tag{6.21}$$

with

$$z_{i-1/2} = (i-1)\Delta z. \tag{6.22}$$

Denote the approximation of the unknown c in Ω_i by $y(z_i, \tau)$. Then, by using the way proposed in [Qamar *et al.* (2006)], we have

$$y(z_i, \tau) = \frac{1}{\Delta z} \int_{z_{i-1/2}}^{z_{i+1/2}} c(z, \tau) dz, \ i = 1, \cdots, N \tag{6.23}$$

Substituting Equation (6.23) into Equation (6.13) yields

$$\begin{aligned} \frac{dy(z_i, \tau)}{d\tau} = & \frac{1}{(\alpha + k)\Delta z Pe} \left(\frac{\partial y}{\partial z} \Big|_{z_{i+1/2}} - \frac{\partial y}{\partial z} \Big|_{z_{i-1/2}} \right) \\ & - \frac{1}{\Delta z} \left(y(z_{i+1/2,\tau}) - y(z_{i-1/2}, \tau) \right). \end{aligned} \tag{6.24}$$

The intermediate functions in Equation (6.24) can be expressed using the κ-flux interpolation scheme. For $\kappa \in [0, 1]$ and $i = 2, \cdots, N-1$, it follows from Section 2.8 of Chapter 2 that

$$\frac{\partial y}{\partial z}\Big|_{z_{i+1/2}} = \frac{y(z_{i+1}, \tau) - y(z_i, \tau)}{\Delta z}, \tag{6.25}$$

$$\frac{\partial y}{\partial z}\Big|_{z_{i-1/2}} = \frac{y(z_i, \tau) - y(z_{i-1}, \tau)}{\Delta z}, \tag{6.26}$$

and

$$y(z_{i+1/2},\tau) = y(z_i,\tau) + \frac{1+\kappa}{4}(y(z_{i+1},\tau) - y(z_i,\tau)) + \frac{1-\kappa}{4}(y(z_i,\tau) - y(z_{i-1},\tau)), \tag{6.27}$$

$$y(z_{i-1/2},\tau) = y(z_{i-1},\tau) + \frac{1+\kappa}{4}(y(z_i,\tau) - y(z_{i-1},\tau)) + \frac{1-\kappa}{4}(y(z_{i-1},\tau) - y(z_{i-2},\tau)). \tag{6.28}$$

We also have

$$y(z_{1/2},\tau) = c(0,\tau), \tag{6.29}$$

$$y(z_{1/2},\tau) = \frac{y(z_1,\tau) + y(z_2,\tau)}{2}, \tag{6.30}$$

$$y(z_{N+1/2},\tau) = y(z_N,\tau) + \frac{y(z_N,\tau) - y(z_{N-1},\tau)}{2}, \tag{6.31}$$

and

$$\frac{\partial y}{\partial z}\Big|_{z_{1/2}} = \frac{-8c(0,\tau) + 9y(z_1,\tau) - y(z_2,\tau)}{3\Delta z}, \tag{6.32}$$

$$\frac{\partial y}{\partial z}\Big|_{z_{N+1/2}} = \frac{y(z_N,\tau) - y(z_{N-1},\tau)}{\Delta z}. \tag{6.33}$$

Applying Equation (6.24) and evaluating it at the mesh points yield a set of ODEs. The resulting ODEs can be solved by using any of time integrator schemes.

6.5 The Alexander Method for Time Integration

The Alexander method is a time integrator, which can be used for numerical computation of ODEs. It is a third-order semi-implicit Runge-Kutta method, and requires several implicit equations to be solved in series. The Alexander method has 3 stage equations with the order of 3. The stability of this method is L-stable, meaning that it not only has the whole negative half plane as stability regions but also be capable of dampening out possible oscillations in the numerical solution even for very large negative eigenvalues. The good stability of the method makes it particularly well-suited for stiff problems as in our cases studies for column chromatography.

The Butcher block of the Alexander method looks like [Yao *et al.* (1998)]:

$$
\begin{array}{c|ccc}
\eta & \eta & 0 & 0 \\
\phi & \phi-\eta & \eta & 0 \\
1 & b_1 & b_2 & \eta \\
\hline
1 & b_1 & b_2 & \eta
\end{array}
$$

where

$$\eta = 0.4358665, \phi = (1+\eta)/2,$$
$$b_1 = -(6\eta^2 - 16\eta + 1)/4, b_2 = (6\eta^2 - 20\eta + 5)/4.$$

The information contained in this block can be translated by the following representations. For simplicity, we define the following general form of ODE (6.24) obtained from spatial discretization. For a specific point along the column length z_i, the ODE becomes:

$$\frac{dy(z_i,\tau)}{d\tau} = f(\tau, y(z_i,\tau))$$

Discretize along time axis with step size of h, and denote $y_{i,j} = y(z_i, \tau_j)$. The approximation of $c(z_i, \tau_{j+1})$ at τ_{j+1} can be calculated through the following procedure. The stage times are:

$$\begin{cases} \tau_{j,1} = \tau_j + \eta h, \\ \tau_{j,2} = \tau_j + \phi h, \\ \tau_{j,3} = \tau_j + h. \end{cases} \tag{6.34}$$

The stage values are calculated from

$$\begin{cases} y^1_{i,j} = y_{i,j} + \eta h K_1, K_1 = f(\tau_{j,1}, y^1_{i,j}), \\ y^2_{i,j} = y_{i,j} + (\phi-\eta) h K_1 + \eta h K_2, K_2 = f(\tau_{j,2}, y^2_{i,j}), \\ y^3_{i,j} = y_{i,j} + b_1 h K_1 + b_2 h K_2 + \eta h K_3, K_3 = f(\tau_{j,3}, y^3_{i,j}). \end{cases} \tag{6.35}$$

The overall slope estimation is

$$K^* = b_1 K_1 + b_2 K_2 + \eta K_3. \tag{6.36}$$

The new point for the next time step is given by

$$y_{i,j+1} = y_{i,j} + hK^*. \tag{6.37}$$

It is seen from Equation (6.35) that the stage value calculations are conducted through implicit equations. Thus, Newton's iteration method will be used as an implicit equation solver to estimate each of the stage values. Let m denote the Newton's interation step. The first stage estimation in the Newton iteration is

$$y^{1,[m+1]}_{i,j} = y^{1,[m]}_{i,j} - \left(1 - \eta h \frac{\partial f}{\partial y}\Big|_{y^{1,[m]}_{i,j}}\right)^{-1} \left(y^{1,[m]}_{i,j} - y^1_{i,j} - \eta h f(\tau_j, y^{1,[m]}_{i,j})\right). \tag{6.38}$$

Then, the initial condition of the column chromatographic process will be used to start the integration.

6.6 Solutions to the Chromatographic Process Model

Analytical solutions to the simplified column chromatographic process model in Equation (6.13) have been shown previously in Figure 6.5 for Péclet Numbers $Pe = 50$ and 500, respectively. Figure 6.5 shows that the the steepness of the wave shape of the solution changes with the change in Péclet Number Pe. Particularly, when Pe is large, the wave shape of the solution becomes very steep.

Numerical computation of the process model should be able to capture the steep changes in the wave shape of the solution. In the implementation of the high resolution method, we set the step size of the spatial discretization to be $\Delta z = 0.01$ and $\Delta z = 0.002$, respectively. Obviously, not only the spatial step size but also the temporal step size will affect the performance of the numerical approximations. However, we put our emphasis on the study of spatial discretization technique in this chapter. The temporal step size is thus set to be $h = 0.002$ for all simulation trials.

With the high resolution method, Tables 6.1 and 6.2 tabulates the numerical results of the process models at $Pe = 50$ and $Pe = 500$, respectively.

Table 6.1 Results at $Pe = 50$ $(\alpha + k = 20, \tau = 10, \tau_0 = 0.4)$.

		$\Delta z = 0.01$		$\Delta z = 0.02$	
z	Analytical Solution	Numerical Solution	Error	Numerical Solution	Error
0.1	2.40E-4	2.82E-4	4.05E-5	2.48E-4	7.32(-6)
0.2	2.73E-3	3.02E-3	2.86E-4	2.78E-3	5.44E-5
0.3	1.39E-2	1.48E-2	9.28E-4	1.41E-2	1.82E-4
0.4	3.79E-2	3.92E-2	1.34E-3	3.81E-2	2.68E-4
0.5	5.80E-2	5.84E-2	3.86E-4	5.81E-2	8.51E-5
0.6	5.12E-2	5.02E-2	-1.05E-3	5.07E-2	-2.03E-4
0.7	2.64E-2	2.68E-2	1.22E-3	2.72E-2	2.46E-4
0.8	8.04E-3	8.43E-3	5.61E-4	8.82E-3	1.16E-4
0.9	1.45E-3	1.59E-3	1.30E-4	1.42E-3	-2.81E-5
1.0	1.56E-4	1.13E-4	-4.23E-5	2.26E-4	-2.94E-5

To further demonstrate the power of the high resolution method in dealing with stiff problems, Table 6.3 gives a comparison of the high resolution method with upwind-1 finite difference method. The same integrator has been uses in all these numerical methods. For the case of $Pe = 500$, we also graphically present comparisons of numerical solutions

Table 6.2 Results at $Pe = 500$ ($\alpha + k = 20, \tau = 10, \tau_0 = 0.4$).

		$\Delta z = 0.01$		$\Delta z = 0.02$	
z	Analytical Solution*	Numerical Solution	Error	Numerical Solution	Error
0.2	-	-	-	-	-
0.3	1.23E-5	-3.30E-6	-1.56E-5	1.35E-5	1.27E-6
0.4	1.92E-2	2.55E-2	6.31E-3	2.01E-2	9.47E-4
0.5	1.78E-1	1.69E-1	-8.29E-3	1.77E-1	-6.60E-4
0.6	1.05E-2	8.30E-3	-2.24E-3	1.00E-2	-5.38E-4
0.7	4.24E-6	-7.24E-6	-1.15E-5	3.70E-6	-5.43E-7
0.8	-	-	-	-	-

Table 6.3 Performance of discretization methods.

Péclet Number Pe	$Pe = 50$			$Pe = 500$		
Peak Position at $\tau = 10$	$z = 0.528$			$z = 0.492$		
Peak exact solution	0.059139			0.17957		
	Computed Solution	Absolute Error	CPU (sec)	Computed Solution	Absolute Error	CPU (sec)
High Resolution ($N = 100$)	0.059206	4.70E-5	8.2	0.17538	4.19E-3	8.1
High Resolution ($N = 500$)	0.059136	3.00E-6	82.8	0.17933	2.40E-4	83.4
Finite Difference ($N = 500$)	0.057815	1.32E-3	104.5	0.14713	3.24E-2	106.4
Finite Difference ($N = 1000$)	0.058467	6.72E-2	294.5	0.16089	1.89E-2	299.4

The temporal mesh number is 500 for all trials. Simulations are conducted on a personal computer with an Intel's Pentium IV 3.00GHz processor.

along the entire column in Figure 6.6.

As shown in Table 6.3 and Figure 6.6, compared with finite difference methods, the high resolution method achieves higher accuracy with much less computational effort in comparison. Also, the obtained solutions from the high resolution method are free from oscillations. By contrast, finite difference methods with a constant step size is no longer suitable for numerical computation of the PDEs with certain stiffness. They have to incorporate other advanced schemes, e.g., adaptive grid algorithm. However, this will bring extra demand on the computational power.

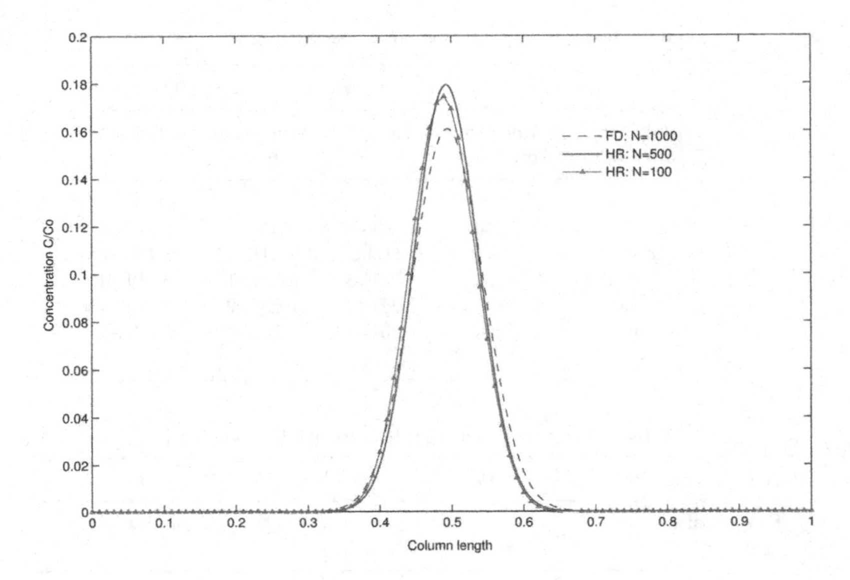

Fig. 6.6 Comparison of numerical solutions using finite difference and high resolution methods.

Another issue worth mentioning is the error distribution plotted in Figure 6.7. The error distributions of the solutions from the high resolution and finite difference methods are quite different. Finite difference produces the largest error at the peak of the wave. For the high resolution method, the error is mostly located at the wave front and tail, but not around the peak. The peak of the wave is actually the turning point of the error sign for the high resolution method. This reflects the capability of the high resolution method in capturing shock or discontinuities of the system.

Summary on High Resolution Methods for Chromatography

In the last few sections, we have explored high resolution discretization techniques for numerical computation of PDEs with singularity. The problem comes from column chromatography. A Transport-Dispersive-Equilibrium linear model has been considered to represent the dynamic behaviour of a single column chromatographic process. The numerical solution to the mathematical model has been used to investigate the numerical power of the proposed methods on the prediction of the transit behaviour of the wave propagation of the system. Comparisons are conducted between the analytical solution and numerical solutions for a

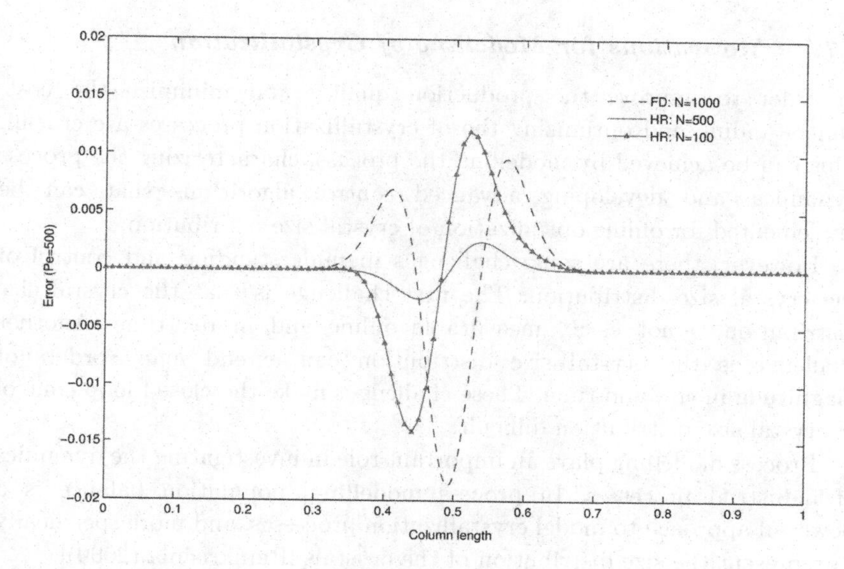

Fig. 6.7 Error distributions at $\tau = 10$.

range of Péclet numbers. The results presented in the last few sections have shown that the high resolution method provides an accurate numerical solution to the PDEs under investigation. The high resolution method is easy to implement, requires less computing effort, and can handle a range of stiff systems.

6.7 Crystallization and Population Balance Equations

Crystallization is one of the important unit operations in chemical engineering. It is widely used in chemical, pharmaceutical, material, semiconductor, and other industries. In this and next few sections, we will discuss crystallization processes, their modelling using population balance equations, model approximations with the high resolution method introduced in Section 2.8, and numerical computation of the process model. Specifically, we will deal with two types of population balance models for crystallization in a batch vessel: one with pure independent growth rate and the other with independent growth rate and nucleation.

6.7.1 *Motivations for Modelling of Crystallization*

In order to improve the production quality and minimize the cost, understanding and optimizing the of crystallization processes are crucial. This can be achieved by modelling the process, characterizing the process dynamics, and developing advanced control algorithms that can be implemented for online optimization of crystal size distribution.

However, there are some challenges in understanding and control of the crystal size distribution. The first challenge is that the crystal size distribution is not easily measurable online and in real-time. Another challenge is the crystal size distribution can extend many orders of magnitude in size and time. These challenges make the closed-loop control of crystal size distribution difficult.

Process modelling plays an important role in investigating the dynamics of industrial processes. In process modelling, population balance is a powerful approach to model crystallization processes, and more specifically to represent the size distribution of the crystals [Ramkrishna (2000)].

Mathematical models including population balance are a class of PDEs, integrodifferential equations or combination of both [Ramkrishna (2000)]. Due to the complexity arising from tracking the change in particle size distribution as particles are born, die, grow or leave a given control volume, deriving analytical solutions are not possible except for a limited number of simple problems. This has motivated much effort in developing numerical computation of population balance equations.

6.7.2 *Development of Population Balance Equation Model*

The population balance model is developed by carrying out a balance over the population density in a fixed subregion of the particle phase space [Liu (2001)]. It can be described by the following general form [Randolph & Larson (1988); Qamar *et al.* (2006)]:

$$\frac{\partial n(L,t)}{\partial t} + \frac{\partial G(L,t)n(L,t)}{\partial L} = q(L,t,n) \tag{6.39}$$

where $n(L,t)$ is the number density function, $G(L,t)$ represents the growth rate, L is the size variable, t denotes time, and $q(L,t)$ stands for the creation and/or depletion rate.

The creation and/or depletion rate $q(L,t)$ can include nucleation, growth, agglomeration and breakage. Both the nucleation and growth can be size-independent or size-dependent. For instance, if the nucleation and agglomeration are considered, the right hand-side of model Equation (6.39) can be rewritten as

$$q(L,t,n) = B(L,t) - D(L,t) + B_0\delta(L) \tag{6.40}$$

where B and D are the birth and death rates, respectively, from the agglomeration. Both B and D are a function of the agglomeration kernel. B_0 is the nucleation rate.

If there is no any agglomeration or nucleation occurring in the processes, $q(L,t) = 0$. In this case, model equations (6.39) and (6.40) become are simplified to

$$\frac{\partial n(L,t)}{\partial t} + \frac{\partial G(L,t)n(L,t)}{\partial L} = 0. \tag{6.41}$$

We are also interested in two scenarios of crystal growth. The first scenario is size-independent growth, and the other is characterized by size-independent growth and nucleation.

For size-independent growth, the growth rate $G(L,t)$ becomes a constant. Therefore, model Equation (6.41) is rewritten as

$$\frac{\partial n(x,t)}{\partial t} + G\frac{\partial n(x,t)}{\partial x} = 0 \tag{6.42}$$

with the initial and boundary conditions

$$n(0,t) = 0; \; n(x,0) = \exp(-100(x-1)^2). \tag{6.43}$$

Later in Section 6.8, the high resolution scheme will be employed to numerically solve the model in Equations (6.42) and (6.43) for crystallization with pure size-independent growth.

In the case of size-independent growth, if nucleation also occurs in the crystallization process, the process model in Equations (6.39) and (6.40) becomes

$$\frac{\partial n(L,t)}{\partial t} + \frac{\partial G(L,t)n(L,t)}{\partial L} = B_0\delta(L). \tag{6.44}$$

This model equation can be further simplified as

$$\frac{\partial n(x,t)}{\partial t} + G\frac{\partial n(x,t)}{\partial x} = 0 \tag{6.45}$$

with the initial and boundary conditions

$$n(x,0) = 0, \quad n(0,t) = \mathbf{1}(t) \tag{6.46}$$

and

$$G = B_0 = 1; \quad \mathbf{1}(t) \;\; \text{is the unit step function.} \tag{6.47}$$

Section 6.9 will be devoted to high resolution numerical computation of the process model in Equations (6.45), (6.46) and (6.47) for crystallization with size-independent growth and nucleation.

6.8 Crystallization with Pure Size-Independent Growth

In this section, the HR numerical scheme will be employed to solve the population balance model with pure size-independent growth.

The model equations for crystallization with pure size-independent growth have been developed in the last section, and are expressed in Equations (6.42) and (6.43). For this specific process model, an analytical solution has been found [Liu (2001)] as

$$n(x,t) = \exp(-100(x - Gt - 1)^2). \tag{6.48}$$

Results from numerical computation will be compared with this analytical solution for performance evaluation.

In the rest part of this section, the constant growth rate G is taken as 3 and the particle size x is taken from 0 to 5 units, i.e.,

$$G = 3, x \in [0, 5].$$

Partition the interval of the particle size $x \in [0,5]$ into N subintervals. The ith subinterval is denoted by $\Omega_i = [x_{i-1/2}, x_{i+1/2}], i = 1, \cdots, N$. The partitioning is conducted such that

$$0 = x_{1/2} < x_1 < x_{1+1/2} < \cdots < x_N < x_{N+1/2} = 5$$

where x_i is the mid-point of the subinterval $\Omega_i = [x_{i-1/2}, x_{i+1/2}], i = 1, \cdots, N$. Then, denote the number density function at x_i, n_i by

$$n_i(t) = \int_{\Omega_i} n(x,t)dx. \tag{6.49}$$

Therefore, the PDE model in Equation (6.42) is reduced to a set of coupled ODEs as

$$\frac{dn_i(t)}{dt} - G\frac{1}{\Delta x_i}(n_{i+1/2}(t) - n_{i-1/2}(t)) = 0, \; i = 1, \cdots, N. \tag{6.50}$$

6.8.1 *Upwinding Scheme for Approximating $n_{i+1/2}$*

If the upwinding scheme is used, then from Equation (2.80) we have

$$n_{i+1/2}(t) = n_i(t) \quad \text{for} \quad i = 1, \cdots, N \tag{6.51}$$

and from the boundary condition (6.43) we have

$$n_{1/2}(t) = n(0,t) = 0. \tag{6.52}$$

Substituting Equations (6.51) and (6.52) into (6.50) yields a set of ODEs as

$$\frac{dn_i(t)}{dt} = G\frac{1}{\Delta x_i}(n_i(t) - n_{i-1}(t)), \; i = 1, \cdots, N \tag{6.53}$$

with the initial condition specified by

$$n_i(0) = n(x_i, 0) = \exp(-100(x_i - 1)^2). \tag{6.54}$$

Equation (6.53) together with initial condition (6.54) can be solved by ODE solvers, such as the Runge-Kutta method mentioned in Section (2.5). Figure 6.8 shows both numerical and analytical solutions of the process model for $N = 200$.

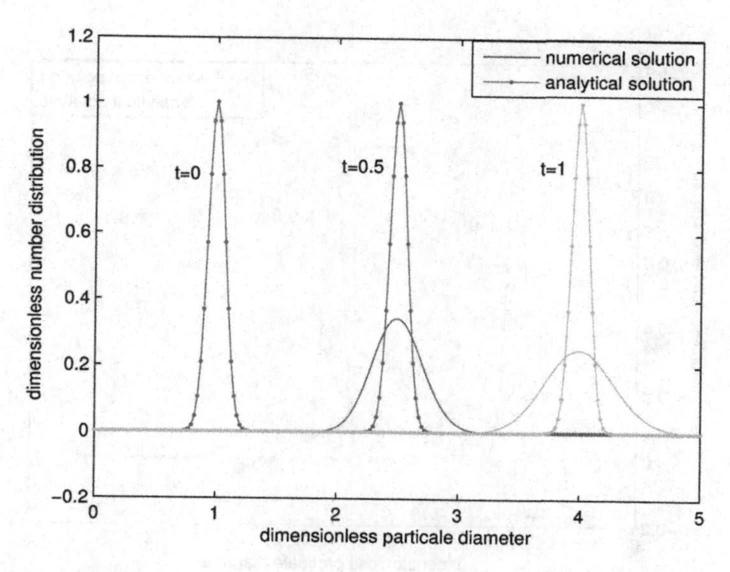

Fig. 6.8 Results of model with pure independent growth using upwind scheme with $N = 200$.

6.8.2 *Approximating $n_{i+1/2}$ via κ-flux Interpolation*

If the κ-flux interpolation scheme is used, then from Equation (2.81) we have

$$n_{i+1/2}(t) = n_i(t) + \frac{1+\kappa}{4}(n_{i+1}(t) - n_i(t)) + \frac{1-\kappa}{4}(n_i(t) - n_{i-1}(t)),$$
$$i = 2, \cdots, N-1 \tag{6.55}$$

As there are no definitions for n_0 and n_{N+1}, we use the 1-flux and -1-flux interpolation schemes to deal with $n_{i+1/2}$ at the boundary, respectively. We have

$$n_{1/2}(t) = n(0,t) + \frac{1}{2}(n_2(t) - n_1(t)),$$
$$n_{N+1/2}(t) = n_N(t) + \frac{1}{2}(n_N(t) - n_{N-1}(t)). \tag{6.56}$$

Combining Equations (6.53), (6.55) and (6.56) gives a set of ODEs with the initial condition specified in Equation (6.54). Simulation of the resulting model is illustrated in figure (6.9) for $N = 200$ and $\kappa = 1/3$ in Equation (6.55).

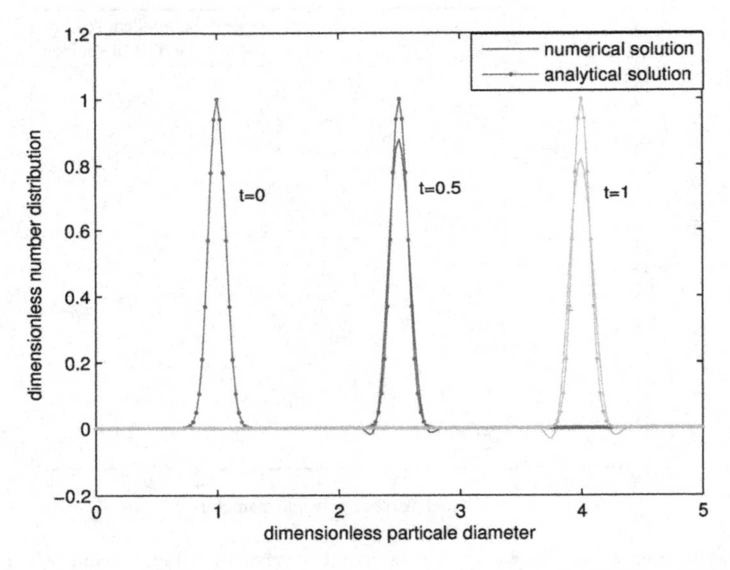

Fig. 6.9 Results of model with pure independent growth using 1/3-flux interpolation scheme with $N = 200$.

6.8.3 *Optimized 1/3-flux Interpolation for Approximating $n_{i+1/2}$*

If taking $\kappa = 1/3$, then from Equation (6.55) we have

$$n_{i+1/2}(t) = n_i(t) + \frac{1}{3}(n_{i+1}(t) - n_i(t)) + \frac{1}{6}(n_i(t) - n_{i-1}(t)),$$
$$i = 2, \cdots, N-1 \tag{6.57}$$

Let us introduce a monitor defined by

$$r_i^+ = \frac{n_{i+1} - n_i + \epsilon}{n_i - n_{i-1} + \epsilon}, \tag{6.58}$$

which is the ratio of consecutive solution gradients. Combining Equations (6.57) and (6.58) and considering Equation (2.83) lead to the optimized flux interpolation scheme as described by

$$n_{i+1/2}(t) = n_i(t) + \frac{1}{2}\Phi(r_i^+)(n_i(t) - n_{i-1}(t)). \tag{6.59}$$

Simulation results of the model with the optimized flux interpolation are shown in Figure 6.10 for $N = 200$ and $\epsilon = 10^{-10}$.

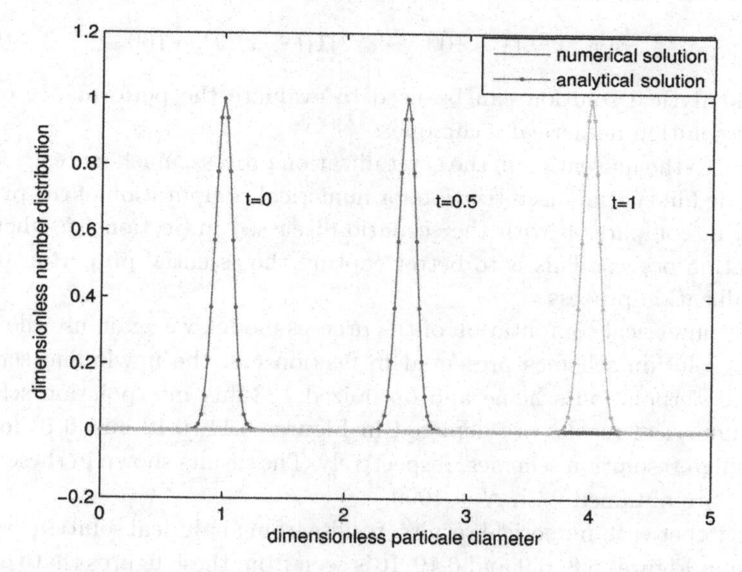

Fig. 6.10 Results of model with pure independent growth using optimized 1/3-flux interpolation scheme with $N = 200$.

It is clearly seen from Figures 6.8, 6.9 and 6.10 that the optimized scheme gives the best computing results and the upwinding scheme behaves the worst. This phenomenon can be interpreted through analytical analysis. The upwinding scheme gives only the first-order accurate approximation, while the flux interpolation schemes give a higher-order accurate approximations.

6.9 Process with Size-Independent Growth and Nucleation

In Section 6.8, numerical computation for crystallization process model has been discussed for the scenario with size-independent growth. This section investigates another scenario with nucleation in addition to size-independent growth. It will show how the high resolution method can be applied to this scenario for numerical computation of the process model.

The mathematical model for crystallization processes with size-independent growth and nucleation has been developed previously in Section 6.7.2. It is expressed in Equation (6.44), which is further simplified to Equations (6.45) and (6.46) and (6.47).

For this model, an analytic solution is available, which is given by

$$n(x,t) = (1 - 100^{-P(t-x)})\mathbf{1}(t-x), P = 100. \tag{6.60}$$

This analytical solution can be used to evaluate the performance of the high resolution numerical technique.

Due to the nucleation in the crystallization process, much more grids will be needed in spatial discretization for numerical computation of the process model in comparison with the scenario discussed in Section 6.8 where no nucleation occurs. This is to better capture the essential properties of the crystallization process.

For numerical computation of the process model, we again use the same high resolution schemes presented in Section 6.8: the upwinding scheme, κ-flux interpolation scheme and optimized 1/3-flux interpolation scheme. Our numerical results are depicted in Figures 6.11, 6.12 and 6.13 for the three high resolution schemes, respectively. The results shown in these three figures are obtained with $N = 1000$.

Together with numerical results, results from analytical solution are also shown in Figures 6.8, 6.9 and 6.10. It is seen from these figures that the two interpolation schemes outperform the upwinding scheme. This is consistent with what we have seen in Section 6.8. The optimized 1/3-flux interpolation scheme behaves slightly better than the 1/3-flux interpolation scheme for this example. However, the differences between these two schemes are not visible in Figures 6.9 and 6.10.

Summary on High Resolution Methods for Crystallization

The last few sections have presented high resolution numerical methods for numerical computation of mathematical models for crystallization processes. The mathematical models are expressed in PDEs, which are further developed for two specific scenarios: one with pure size-independent growth of crystals and the other with nucleation in addition to size-independent growth. The model problems from these two scenarios are

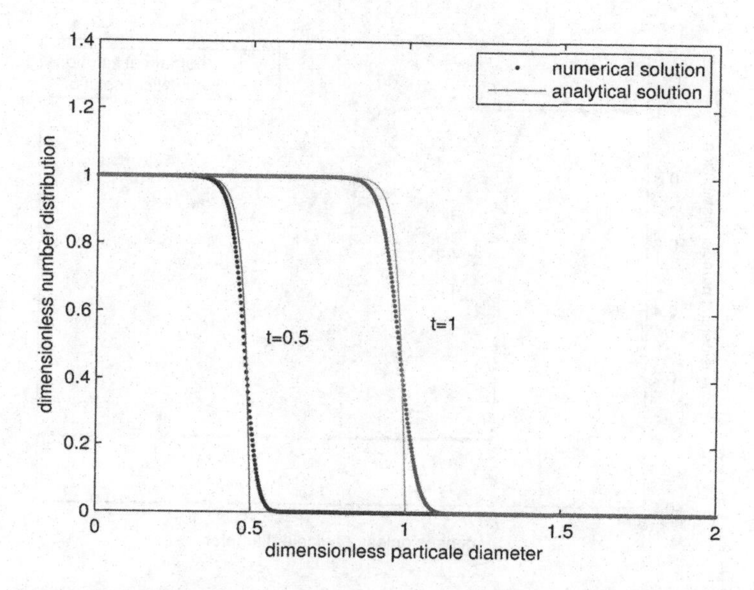

Fig. 6.11 Results of model of Independent growth and Nucleation using upwind scheme with $N = 1000$.

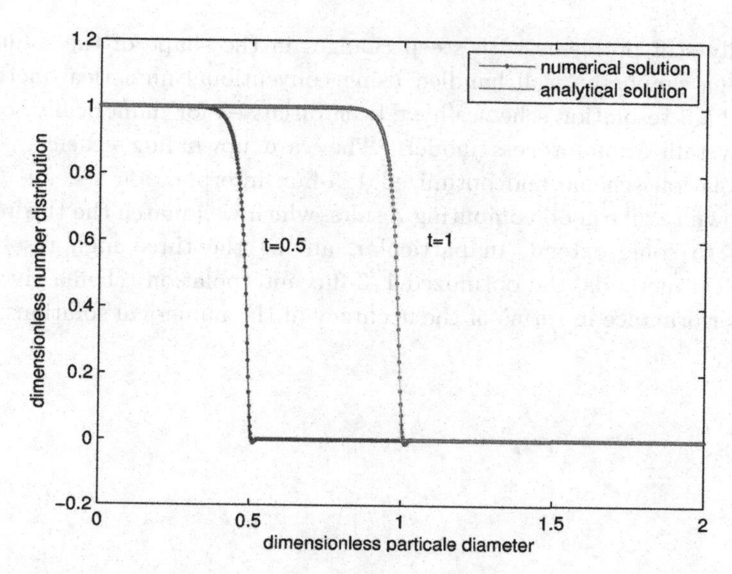

Fig. 6.12 Results of model of Independent growth and Nucleation using 1/3-flux interpolation scheme with $N = 1000$.

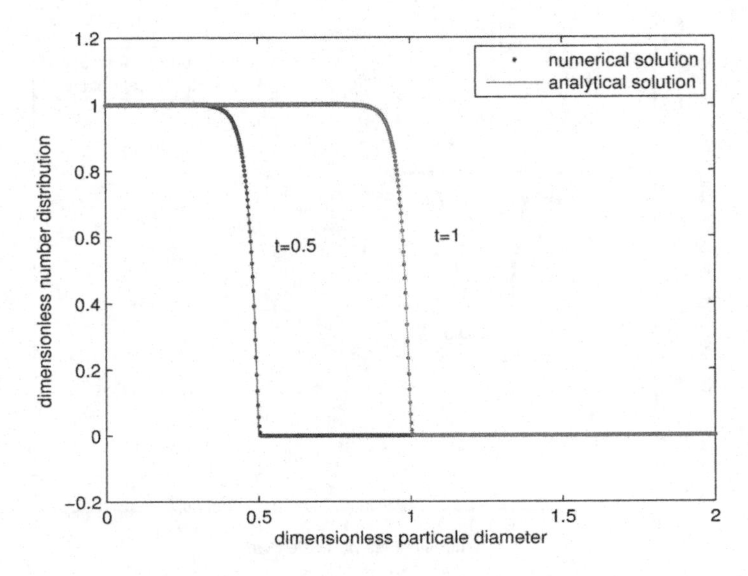

Fig. 6.13 Results of model of Independent growth and Nucleation using optimized 1/3-flux interpolation scheme with $N = 1000$.

typically stiff problems with steep changes in the shape of the solutions, and thus cannot be well handled using conventional numerical methods. Three high resolution schemes have been discussed for numerically solving the crystallization process models. They are upwinding scheme, κ-flux Interpolation scheme and optimized 1/3-flux interpolation scheme. They are shown to give good computing results, which well match the theoretical results to some extend. In particular, among the three high resolution numerical methods, the optimized 1/3-flux interpolation scheme gives the best performance in terms of the accuracy of the numerical solutions.

Chapter 7

Comparative Studies of Numerical Methods for SMB Chromatographic Processes

Several widely used types of numerical methods have been discussed so far in previous chapters for computation of mathematical models of complex industrial processes. Each of them has its own advantages and drawbacks, and is thus suitable for specific applications or systems under specific operating conditions. Therefore, a good understanding of the suitability of the numerical methods in specific systems is crucial for industrial applications.

This chapter conducts comprehensive comparative studies of various numerical methods for complex processes of simulated moving bed (SMB) chromatography, which represent a type of separation processes in chemical engineering. The SMC chromatography (SMBC) technique is used to separate particles and/or chemical compounds that would be difficult or impossible to resolve otherwise. It exhibits non-linear, non-ideal, multivariable-coupling, and hybrid system dynamics with significant time delay. Given the value of the products and operating costs of SMBC processes, the effort in SMBCP modelling has big impact for process industries, and will be translated into significant monetary benefits. However, developing techniques for solving SMBCP models is challenging due to the high order model equations with complexity.

This chapter will firstly build a fundamental knowledge of SMBC processes through a brief introduction to the operation and advantages of SMBC separation. Then, the challenges in seeking numerical solutions to SMB models are identified. This is followed by mathematical modelling od SMBC processes. Various numerical computing methods to be studied in this chapter are also presented. After that, comparative studies of these numerical methods are conducted through two SMBC applications.

Much of the content of this chapter are taken from our previous publications [Yao *et al.* (1998, 2010a,b); Zhang *et al.* (2008)].

7.1 Chromatographic Separation Processes

In order to have a good understanding of the mechanisms of SMBC processes, we will give a technical introduction to the basic operating methods in chromatographic separation operation. There are three basic types of operating methods for chromatographic separation processes: fixed-bed chromatography, moving-bed chromatography and the simulated moving-bed chromatography. Following this technical introduction are discussions on economical advantages of SMBC. Challenges in seeking numerical solutions to SMBC process models will also identified.

7.1.1 *Fixed Bed Chromatography*

Fixed-bed chromatography involves a column packed with adsorbent and then filled with liquid desorbent. A fixed amount of mixed sample is fed in at one end of the column. If we continually feed in liquid desorbent, each component of the sample will move through the adsorbent layers at migration rates determined by their individual adsorption affinity. During this movement, they will separate from each other. By collecting the liquids as they successively drip out of the column, the mixture can be separated into fractions each abundant with one of the component materials.

Fixed-bed adsorption systems are typically operated in a cyclic manner with each bed repeatedly undergoing a sequence of steps, such as pressurization, adsorption, blowdown, and desorption. Since the types and densities of the desorbent can be easily changed, fixed-bed chromatography can be applied to the separation of many types of useful substances.

However, the fixed-bed chromatographic process results in a material with a low concentration, in cases where a more precise separation is required, higher costs are required in raising the material's concentration.

There are some additional deficiencies when fixed bed method is used on a large scale:

- The entire adsorbent bed is not efficiently utilized;
- A large quantity of desorbent is consumed, and the separated components are obtained in a diluted state;

- In order to obtain a successful separation, a sufficiently large difference in the adsorption affinities of the adsorbates is required; and
- The operation is not continuous.

Due to these deficiencies, there have been many innovative attempts to improve the fixed bed operation so that it can be used on an industrial scale. The moving bed method is one of the ways to make the fixed-bed system continuous.

7.1.2 *Moving Bed Chromatography*

Moving bed chromatography is graphically illustrated in Figure 7.1. As shown in Figure 7.1, in moving bed chromatography, the desorbent (liquid phase) is continuously fed into one end of the column, while the adsorbent (solid phase) is made to move in the opposite direction. The feed mixture is supplied at the middle of the bed.

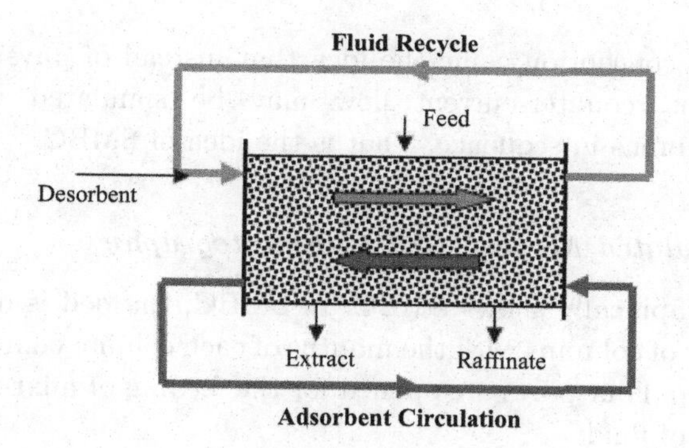

Fig. 7.1 Moving bed chromatography.

In moving bed operation, starting from the point where the feed mixture is fed, components A and B move in opposite directions from one another within the column. If the mixture is continuously fed into such a moving-bed column, components A and B will be continuously separated to both ends of the column.

The benefits from a moving bed are less desorbent as well as adsorbent than those used in fix bed chromatography. However, there are some unusual characteristics in moving-bed chromatography, and these characteristics cannot be accomplished with batch chromatography. For example, there is a useful path length effect in moving-bed operation. A longer path length will generally improve the chromatographic separation because additional equilibrium steps, or 'theoretical plates' become available. With the moving bed technique, the path length of adsorbent, which interacts with the feed mixture, can be increased beyond the length of the adsorbent in the system by increasing its internal flow rate. Increasing internal flows in both directions means that the entering mixture will contact with a much longer path of resin before exiting. Instead of being constrained to path length as in batch chromatography, the moving bed system can increase path length without increasing the amount of adsorbent used. The adsorbent can be regenerated as soon as its role in the adsorption step has been completed.

However, in a large industrial device, it is extremely difficult to uniformly move the adsorbent without disturbing the adsorption band. Moving the column is mechanically complex. The equipment required will inevitably be more expensive and need to cope with attrition of the adsorbent.

From this conception came the idea that instead of physically moving the adsorbent, counter-current flow may be simulated by switching adsorbent columns in sequence. That is the idea of SMBC.

7.1.3 *Simulated Moving Bed Chromatography*

Figure 7.2 graphically shows SMBC. In SMBC, the bed is divided into a small number of columns with the mouths of each column connected to form a circular loop. Four ports are opened for the feeding of mixture, desorbent and drawing of fluids.

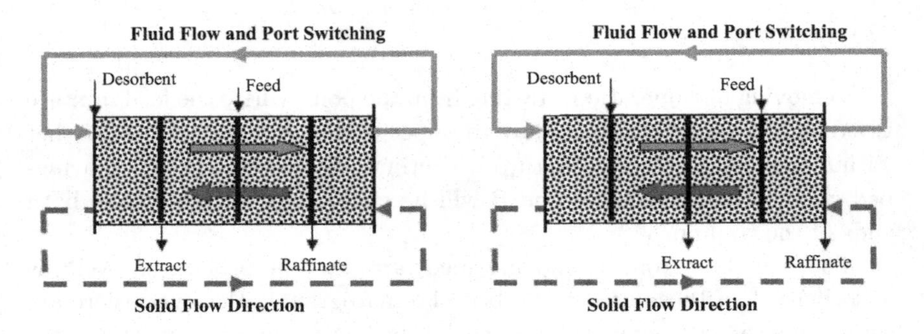

Fig. 7.2 Simulated moving bed chromatography.

The following describes how an SMBC separation system works. While the fluids are circulating inside the column, assume that feed mixture F, desorbent D, extract E, and raffinate R are allowed to enter or leave the column continuously from each of the openings. Then, successively for the length of one full column at a time, the positions of the openings for F, D, E, R are changed in the direction of the circular flow at a regularly fixed time point. By operating the device under these conditions, it seems as if the adsorbent moved in the opposite direction to the flow of the fluids. Therefore, components A and B move in the opposite direction from the introduction point of the feed mixture, and each component can be removed in a continuous manner from its respective withdrawal points E and R.

A well known application of this principle is the SORBEX system, where a rotatory valve controls the motion of the ports along the column. The same result can be achieved by connecting each section of the column with many on-off valves, each of which is in turn a feed point or draw off point or simple connection depending on the particular role of the section at a particular time. Typically, four-section cascade SMB as shown in Figure 7.3 is the usual process scheme adopted in commercial applications and academic research.

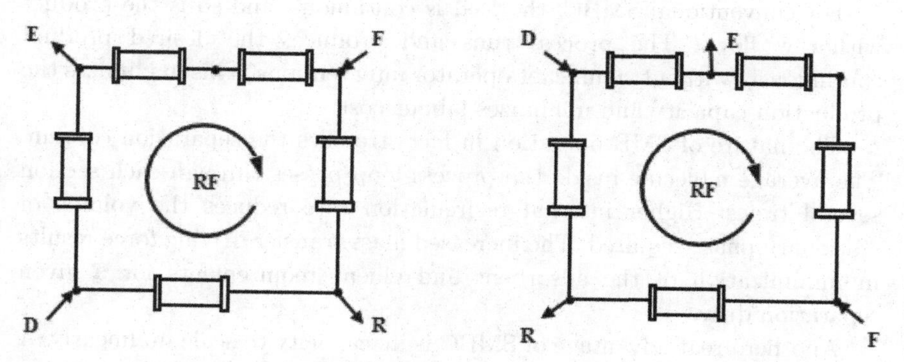

Fig. 7.3 An SMB unit with 5 columns in a 4-section configuration of 1/2/1/1 (RF: direction of fluid flow and port switching).

SMB operation is not a perfect simulation of true moving bed (TMB) operation. The adsorbent movement simulation by valve switching is not accomplished in a continuous manner, but rather, jumps intermittently between cells. This means that as more cells are added to an SMB, a true moving bed is closely simulated. An infinite number of cells and infinitely short switching periods would be needed for an exact simulation of true moving bed operation. With sharp mass transfer zones or when high purity products are not required, one column per zone is satisfactory. When sharper separations are required, several segments per zone are needed. 4 to 12 columns are typically enough to obtain most of the benefits of true moving bed operation.

7.1.4 *Economical Advantages of SMBC Operation*

SMBC processes provide the advantages of a continuous counter-current unit operation while avoiding the technical problems of a true moving bed. Economical advantages of SMBC operation are described in the following.

An SMBC process can use a higher concentration of solute in the feed than a batch process and the total concentration of products are much higher. By reducing the required volumes of stationary and mobile phase, SMBC processes can achieve higher productivity with lower solvent consumption, implying that the products are less diluted and consequently the product recovery step is easier and cheaper. The savings in solvent consumption from using an SMBC process typically range from a factor of 3-10 when compared to batch chromatography [Pynnonen (1998)].

For conventional SMBC, the feed is continuous, and so is the product withdraw flows. The process runs and produces the desired product continuously without significant operator interventions. This maximizes the production capacity and minimizes labour cost.

The nature of SMB operation in fact stretches the separation column. The average molecule inside the process loop passes through each section several times. Higher internal recirculation rate reduces the volume of stationary phase required. The increased mass transfer driving force results in minimization of the adsorbent and eluent requirements for a given separation duty.

Another great advantage of SMBC is its capacity to scale up linearly. A pilot scale process can be reproduced quickly at production scale without sacrificing the purity and production rate.

The SMB technique was first developed to separate isomeric mixtures such as xylenes and sugars using resins or zeolites. Nowadays, it is widely and successfully used in various areas. In petrochemical industry, it is used for separation of xylene isomers [Jin and Wankat (2005); Minceva and Rodrigues (2002)]). In pharmaceutical engineering, it is adopted for separation of enantiomers [Schulte (2001); Wang and Ching (2004); Wongso (2004)]. In food industries, it is designed for separation of sugars [Azevedo

& Rodrigues (2000); Lu *et al.* (2006)] or amino acid [Molnar *et al.* (2005); Xie *et al.* (2005)]. In addition, it has also been used in fine chemical and other industries.

7.1.5 *Challenges in Solving SMBC Process Models*

Industrial applications of the SMBC technology require a good design of the SMB systems and effective control of the system operation. It is an emergent requirement to improve the SMBC process operation for higher product quality, productivity, efficiency, and robustness.

Optimization of the SMBC operation can be achieved through optimizing a performance index (or multiple indices) with respect to manipulated process variables. It requires accurate process models as constraints and a powerful numerical solver for complex model solutions. In addition, process modelling is also necessary for scale-up from laboratory to industrial sale, for prediction of process dynamics, and for on-line and real-time process control.

However, for SMBC process, we encounter at least the following two challenges: process modelling and model computation.

An adsorption column has several features that make the modelling task particularly challenging. These include: strong nonlinearities in the adsorption equilibrium isotherms, interference effects due to the competition of solutes for adsorbent sites, mass transfer resistances between the fluid and solid phases, and fluid-dynamic dispersion phenomena. The interplay of these effects produces steep concentration fronts, which move along the column during the process. This particular property, if accounted for by the mathematical model, constitutes a serious difficulty for the solution procedure.

The difficulty in model computation comes from the complexity of the mathematical model of the SMBC process. As far as a SMB system is concerned, the model consists of a set of PDEs for mass balance over the column, ODEs for parabolic intraparticle concentration profile, and algebraic equations (AEs) for equilibrium isotherms and node mass balances. These equations are highly coupled, making it very difficult, if not impossible, to be solved analytically. Finite discretization of the PDEs gives rise to dynamic systems of a very high order. Problems with sharp variations of solutions require even larger discretization models. The CPU time required for the simulation of an SMBC process largely depends on how many equations need to be solved. Systems with more components and stiff concentration profiles require more equations. The large size is a problem due to its intensive computation when on-line optimization and real-time control are necessary.

These challenges leave us the following open questions: *Which numerical solution technique is more suitable to circumvent the steep change encountered in chromatographic separation while keeping the computational demand reasonably low for real-time control?*

This chapter will provide a practically useful tool for achieving numerical model solutions. Three spatial discretization methods are chosen for this particular system. They are upwind-1 finite difference, wavelet-collocation method, and high resolution method. Comparative studies are carried out through two cases:

- A glucose-fructose separation process is firstly chosen with its relatively simple isotherm representations. Simulations are conducted using both wavelet collocation and upwind finite difference methods. Quantitative evaluations are made on product purity and yield. Computer elapsed time is compared with those reported in the open literature.
- For more complicated applications, an enantiomers separation process is selected. As the PDE model presents a certain degree of singularity, the wavelet collocation and high resolution methods will be adopted for spatial discretization. Apart from the computational time, a specifically defined relative error will be used as a criterion to evaluate the convergence performance of proposed algorithms.

7.2 Dynamic Modelling of SMBC Processes

A dynamic SMB model consists of two main parts: a single chromatographic columns model, and the node balance model describing the connection of the columns combined with the cyclic switching. A schematic description of the modelling system for SMBC processes is depicted in Figure 7.4.

The switching operation of an SMBC process can be represented by a shifting of the initial or boundary conditions for the single columns. The stationary regime of the process is cyclic steady-state (CCS), in which an identical transient takes places in each section during each period between two valve switches. Normally, CSS is determined by solving the PDE system repeatedly for each step of the cyclic process in sequence, using the final concentration profile for each step as the initial condition for the next step in the cycle.

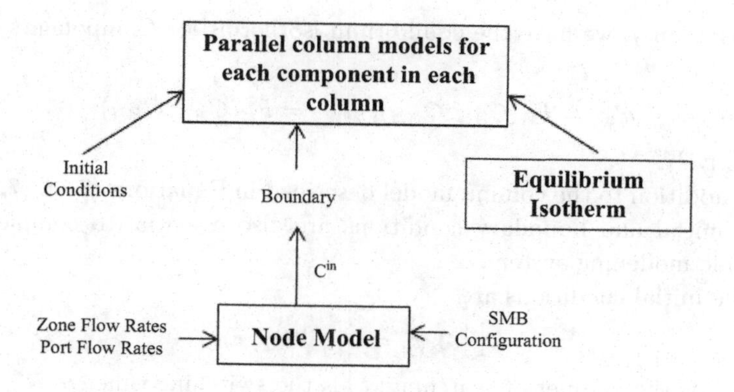

Fig. 7.4 SMB dynamic modelling system.

In the following, detailed model equations are given for SMBC processes.

7.2.1 *Column Model for Chromatography*

The selected column model structure is the same as that described in Section 6.1, i.e., the transport-dispersive-LDF model. We will expand the model to a multi-columns binary separation process. Equation (6.1) for transport-dispersive-LDF model indicates that the mass balance of component i and column j in the mobile phase (or bulk-liquid phase) is represented by

$$\frac{\partial C_{i,j}}{\partial t} + u_j \frac{\partial C_{i,j}}{\partial x} = D_T \frac{\partial^2 C_{i,j}}{\partial x^2} - \frac{1-\epsilon_T}{\epsilon_T}\frac{\partial q_{i,j}}{\partial t}, \tag{7.1}$$

where $C_{i,j}$ stands for liquid phase concentration of component i in column j; $q_{i,j}$ represents solid phase concentration of component i in column j; $i = A, B$; $j = 1, 2, \cdots, N_{column}$; N_{column} denotes the total number of columns; u_i is the interstitial velocity of component i; D_T is a lumped dispersion and diffusion coefficient, and ϵ_T represents total porosity.

Similar to Equation (6.2), a simple Linear Driving Force model is used to represent the overall mass transfer kinetics of component i and column j for the adsorption phase:

$$\frac{\partial q_{i,j}}{\partial t} = k_{\text{eff}}(q^*_{i,j} - q_{i,j}). \tag{7.2}$$

where k_{eff} refers to effective mass transfer resistance, $q*_{i,j}$ is the equilibrium concentration of component i in the interface between two phases in column

j. In column j, we have the equilibrium isotherms for Components A and B as

$$q^*_{A,j} = F_A(C_{A,j}, C_{B,j}),\ q^*_{B,j} = F_B(C_{A,j}, C_{B,j}), \tag{7.3}$$

respectively.

In addition to the column model described in Equations (7.1), (7.2) and (7.3), initial and boundary conditions are also essential to complete the dynamic modelling system.

The initial conditions are

$$C^{[s]}_{i,j}(0, x) = C^{[s-1]}_{i,j}(t_s, x), \tag{7.4}$$

where s is the number of switching; t_s is the switching time.

The boundary conditions are described by

$$\frac{\partial C_{i,j}}{\partial x} = \frac{u_j}{D_T}(C_{i,j} - C^{in}_{i,j}) \quad \text{for } x = 0 \text{ at } t > 0, \tag{7.5}$$

$$\frac{\partial C_{i,j}}{\partial x} = 0 \quad \text{for } x = L \text{ at } t > 0. \tag{7.6}$$

$C^{in}_{i,j}$ are subject to the material balances at the column conjunctions, and will be determined by node model, which will be discussed in the following.

7.2.2 *Node Model for an SMB System*

Consider a 4-section SMB system shown in Figure 7.5. Assume that the dead volume by the switching devices, connecting tubes, and other parts is negligible.

We use the following notations in developing the node model:

$Q_I, Q_{II}, Q_{III}, Q_{IV}$: volumetric flow rates through the corresponding sections I, II, III, and IV, respectively;

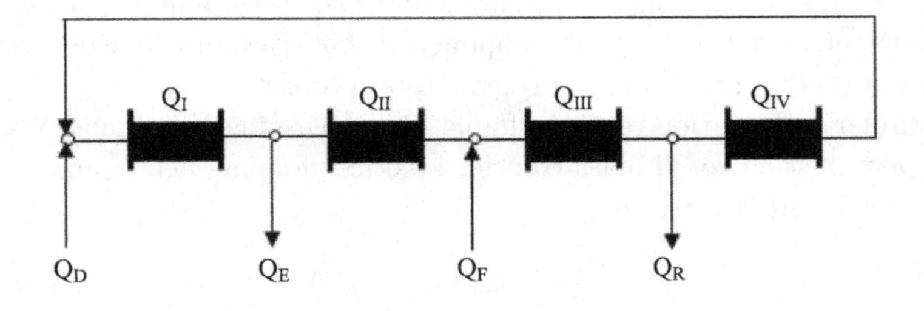

Fig. 7.5 Flow diagram of a 4-section SMB system.

Q_D, Q_F, Q_E, Q_R: the flow rates of desorbent, feed, extract and raffinate, respectively; and

$C_{i,j}^{in}, Q_{i,j}^{out}$: the concentrations of component i at the outlet and the inlet of section j.

The flow and integral mass balance equations at each node are given in the following equations:

Desorbent Node (eluent):

$$Q_I = Q_{IV} + Q_D, \quad C_{i,I}^{in} = C_{i,I}^{out} Q_{IV} + C_{i,D} Q_D. \tag{7.7a}$$

Extract draw-off node:

$$Q_{II} = Q_I - Q_E, \quad C_{i,E} = C_{i,II}^{in} = C_{i,I}^{out}. \tag{7.7b}$$

Feed node:

$$Q_{III} = Q_{II} - Q_F, \quad C_{i,III}^{in} Q_{III} = C_{i,II}^{out} Q_{II} + C_{i,F} Q_F. \tag{7.7c}$$

Raffinate draw-off node:

$$Q_{IV} = Q_{III} - Q_R, \quad C_{i,R} = C_{i,IV}^{in} = C_{i,III}^{out}. \tag{7.7d}$$

Other node:

$$C_{i,j}^{in} = C_{i,j-1}^{out}. \tag{7.7e}$$

Thus, an SMBC model is constructed by all the single column models connected in series by boundary conditions. The dominating parameters of the interstitial velocity u_j and the inlet concentration of each column are constrained by node models. The overall SMBC model is a distributed parameter system, in which the dependent variables vary with axial position and time. The number of model equations is enlarged in folder with the number of columns involved and the components to be separated.

Although for specific cases there exist analytical solutions to a single column model, it is generally impossible to obtain analytical solutions for an SMB model system with all these highly coupled equations. This demands numerical methods for model computation.

7.3 Numerical Computation

As it is almost impossible to solve the SMBC system model analytically, numerical computing of the system model becomes crucial. In our study, we will investigate the potential of wavelet collocation methods for solving

SMB models numerically. The effectiveness of this method will be compared with finite difference and high resolution methods through two case studies.

The spatial discretization of Equation (7.1) and its boundary conditions for column model are performed using finite difference, wavelet collocation and high resolution methods, respectively, to transform PDEs to ordinary equations for each spatial mesh point (or collocation point).

Let $y_{i,j}$ denote the approximation of $C_{i,j}$, for all methods, we have **Initial conditions** describing the status of the column at the beginning of the switching:

$$y_{i,j}^{[s]}(0,x) = y_{i,j}^{[s-1]}(t_s,x). \tag{7.8}$$

For simplicity, in the following expressions we will use y to stand for $y_{i,j}$. Each of the equations developed should be applied to components A and B for N_{column} columns.

7.3.1 *Upwind-1 Finite Difference Discretization*

Discretize the spatial derivatives in Equation (7.1) using finite difference approximations as follows:

$$\frac{\partial y(x_k,t)}{\partial x} = \frac{y(x_k,t) - y(x_{k-1},t)}{\Delta x}$$
$$\frac{\partial^2 y(x_k,t)}{\partial x^2} = \frac{y(x_{k+1},t) - 2y(x_k,t) + y(x_{k-1},t)}{\Delta x^2}$$

This results in an ODE for each spatial mesh point:

$$\begin{aligned}\frac{dy(x_k,t)}{dt} &= \frac{D_T}{\Delta x^2} y(x_{k+1},t) - \left(\frac{u}{\Delta x} + \frac{2D_T}{\Delta x^2}\right) y(x_k,t) \\ &+ \left(\frac{u}{\Delta x} + \frac{D_T}{\Delta x^2}\right) y(x_{k-1},t) \\ &- \frac{1-\epsilon_T}{\epsilon_T} k_{\text{eff}}(q^*(x_k,t) - q(x_k,t)),\end{aligned} \tag{7.9}$$

where $k = 2, \cdots, N_x - 1, \Delta x = x_{k+1} - x_k, N_x$ is the number of mesh points.

This particular problem involves flux boundary conditions represented by Equations (7.5) and (7.6). The boundary conditions are treated using the False Boundary Method. We introduce "fictitious" points x_0 and x_{N_x+1} at both ends of the column.

For $x = 0$, apply a second-order correct approximation for the first derivative:

$$\frac{\partial y(x_1,t)}{\partial x} = \frac{y(x_2,t) - y(x_0,t)}{2\Delta x}.$$

It follows that Equation (7.5) becomes

$$\frac{y(x_2,t)-y(x_0,t)}{2\Delta x}=\frac{u}{D_T}(y(x_1,t)-y^{in}(x_1,t)).$$

Solving for $y(x_0,t)$ gives

$$y(x_0,t)=y(x_2,t)-\frac{2\Delta}{D_T}u(y(x_1,t)-y^{in}(x_1,t)). \tag{7.10}$$

Substituting (7.10) into (7.9) with $k=1$, we have

$$\begin{aligned}\frac{dy(x_1,t)}{dt}&=\frac{D_T}{\Delta x^2}y(x_3,t)-\left(\frac{D_T}{\Delta x^2}+\frac{u^2}{D_T}+\frac{u}{\Delta x}\right)y(x_2,t)\\&+\left(\frac{u}{\Delta x}+\frac{u^2}{D_T}\right)y^{in}(x_1,t)-\frac{1-\epsilon_T}{\epsilon_T}k_{\text{eff}}(q^*(x_1,t)\\&-q(x_1,t)).\end{aligned} \tag{7.11}$$

A similar procedure can be used at $x=L$ for Equation (7.6) in order to obtain

$$y(x_{N_x+1},t)=y(x_{N_x-1},t). \tag{7.12}$$

Substituting into Equation (7.9) with $k=N_x$ yields

$$\begin{aligned}\frac{dy(x_{N_x},t)}{dt}&=\left(\frac{D_T}{\Delta x^2}+\frac{u}{\Delta x}\right)(y(x_{N_x-1},t)-y(x_{N_x},t))\\&-\frac{1-\epsilon_T}{\epsilon_T}k_{\text{eff}}(q^*(x_{N_x},t)-q(x_{N_x},t)).\end{aligned} \tag{7.13}$$

Thus, the finite difference method generates a set of ODE systems with Equations (7.9), (7.11) and (7.13).

7.3.2 *Wavelet Collocation Discretization*

The compactly supported orthonormal wavelets is constructed to approximate the function to the interval of $[0,1]$ through axial coordinate transform $z=x/L$. We have

$$\frac{\partial y(z_k,t)}{\partial z}=\sum_{j=1}^{2^J+1}y(z_j,t)T_{k,j}^{(1)}\quad\text{and}\quad\frac{\partial^2 y(z_k,t)}{\partial z^2}=\sum_{j=1}^{2^J+1}y(z_j,t)T_{k,j}^{(2)}$$

Applying to the system model Equation (7.1), we have

$$\begin{aligned}\frac{dy(z_k,t)}{dt}&=-\frac{u}{L}\sum_{j=0}^{2^j}y(z_j,t)T_{k,j}^{(1)}+\frac{D_T}{L^2}\sum_{j=0}^{2^J}y(z_j,t)T_{k,j}^{(2)}\\&-\frac{1-\epsilon_T}{\epsilon_T}k_{\text{eff}}(q^*(z_k,t)-q(z_k,t)),\end{aligned} \tag{7.14}$$

where $k = 2, 3, \cdots, N-1, N = 2^J + 1$ is the number of collocation points; J is the resolution level; $T_{k,j}^{(1)}$ and $T_{k,j}^{(2)}$ are the first and second derivatives, respectively, for the autocorrelation function of the scaling function.

The left boundary condition at $z = 0$ is

$$\sum_{j=1}^{2^J+1} y(z_j, t)T_{1,j}^{(1)} = \frac{uL}{D_T}(y(z_1, t) - y^{in}(z_1, t)). \tag{7.15}$$

The right boundary condition $z = 1$ is described by

$$\sum_{j=1}^{2^J+1} y(z_j, t)T_{2^J,j}^{(1)} = 0. \tag{7.16}$$

7.3.3 *Discretization using High Resolution Method*

Applying the high resolution method to the PDE model equations of the SMBC process gives the following ODEs:

$$\begin{aligned}\frac{dy(z_k, t)}{dt} &= \frac{D_T}{\Delta x}\left(\frac{\partial y}{\partial x}\Big|_{k+1/2} - \frac{\partial y}{\partial x}\Big|_{k-1/2}\right)\\ &\quad -\frac{u}{\Delta x} y(x_{k+1/2,t} - y(x_{k-1/2}, t))\\ &\quad -\frac{1-\epsilon_T}{\epsilon_T} k_{\text{eff}}(q^*(x_k, t) - q(x_k, t)).\end{aligned} \tag{7.17}$$

The expression of each part in this equation is defined as follows.

$$\begin{aligned}&\frac{\partial y}{\partial x}\Big|_{1/2} = \frac{-8y^{in} + 9y(x_1, t) - y(x_2, t)}{3\Delta x + 8D_T/u},\\ &y(x_{1/2}, t) = \frac{3u\Delta x y^{in} - 9D_T y(x_1, t) + D_T y(x_2, t)}{3u\Delta x + 8D_T},\\ &y(x_{3/2}, t) = \frac{y(x_1, t) + y(x_2, t)}{2},\\ &\frac{\partial y}{\partial x}\Big|_{N_x-1/2} = \frac{y(x_{N_x}, t) - y(x_{N_x-1}, t)}{\Delta x},\\ &\frac{\partial y}{\partial x}\Big|_{N_x+1/2} = 0\\ &y(x_{N_x+1/2}, t) = y(x_{N_x}, t) + \frac{y(x_{N_x}, t) - y(x_{N_x-1}, t)}{2}.\end{aligned}$$

For $k = 2, 3, \cdots, N_z - 1, \kappa \in [0, 1]$, we have

$$\begin{aligned}\frac{\partial y}{\partial x}\Big|_{k+1/2} &= \frac{y(x_{k+1}, t) - y(x_k, t)}{\Delta x},\\ \frac{\partial y}{\partial x}\Big|_{k-1/2} &= \frac{y(x_k, t) - y(x_{k-1}, t)}{\Delta x},\end{aligned}$$

$$
\begin{aligned}
y(x_{k+1/2},t) &= y(x_k,t) + \frac{1+\kappa}{4}(y(x_{k+1},t) - y(x_k,t)) \\
&\quad + \frac{1-\kappa}{4}(y(x_k,t) - y(x_{k-1},t)), \\
y(x_{k-1/2},t) &= y(x_{k-1},t) + \frac{1+\kappa}{4}(y(x_k,t) - y(x_{k-1},t)) \\
&\quad + \frac{1-\kappa}{4}(y(x_{k-1},t) - y(x_{k-2},t)).
\end{aligned}
$$

7.3.4 *Settings for Numerical Computation*

Numerical computation in this chapter has been conducted using Matlab on a personal computer with an Intel's Pentium IV 3.00GHz processor. The IVP integrator is the Alexander Semi-implicit Method (the third-order Runge-Kutta). Mathematical background for this method can be found from Section 6.1. Function "model" returns the expression of ODEs, which are generated from original PDEs through spatial discretization techniques, e.g., finite difference, wavelet collocation and high resolution method.

Using an 8-column SMB bi-separation system as an example, PDEs are defined as: 1-8 for component A in liquid phase; 9-16 for component B in liquid phase; 17-24 for component A in solid phase; and 15-32 for component B in solid phase. As a result of discretization (N mesh points or collocation points for each column), the "model" involves $32 \times N$ ordinary equations. To compromise the large number of ODEs, a big integration step size should be chosen to release the computational pressure. This program has indeed worked very well with a sparse integration step.

It is assumed that the process reaches cyclic steady state after 12 cycles (96 switches) or after the relative error defined by Equation (7.21) is less than 0.001. However, this can always be set up by users according to the requirements for specific applications.

7.4 Case Study I: Fructose-Glucose Separation

The separation of fructose-glucose in de-ionized water used in [Beste *et al.* (2000)] and [Lim (2004)] is taken as one of our case study systems. The SMBC process consists of 8 columns with a configuration of 2:2:2:2. The operating conditions and model parameters are summarized in Table 7.1. The transport-dispersive-LDF model is chosen to represent the column dynamics. Thus, the process model consists of 16 PDEs, 16 ODEs and

Table 7.1 Parameters of the fructose-glucose separation.

Symbol	Value	Symbol	Value (Sp. A-fructose)	Value (Sp. B-fructose)
L(cm)	52.07	$k_{\text{eff},i}(\text{min}^{-1})$	0.72	0.9
D(cm)	2.6	$C_{i,\text{feed}}(g/L)$	363	322
ϵ_T	0.41448			
$t_s(min)$	16.39	$q_A^* = 0.675C_A$		
$Q_F(ml/min)$	1.67	$q_B^* = 0.32C_B + 0.000457C_AC_B$		
	Zone I	Zone II	Zone III	Zone IV
$Q(ml/min)$	15.89	11.0	12.67	9.1
$D_T(cm^2/min)$	1.105	0.765	0.881	0.633

20 AEs connecting all the variables together.

Numerical simulations have been performed in this work using both finite difference and wavelet collocation methods for spatial discretisation. The same integrator is adopted for both methods so that the results can be compared on the effectiveness of different spatial discretization methods.

For the trials of finite difference, N_x, the number of mesh points along one column length, has been chosen as $N_x = 33$ and 65, which are equivalent to the collocation points generated by wavelet level of $J = 5$ and $J = 6$.

Simulations using wavelet collocation are conducted on the level $J = 4, 5$ and 6. The boundary conditions are treated using polynomial interpolation proposed by [Liu *et al.* (2000)] with the degree M of 1, 2 and 3. The number of mesh points along the time axis is 5 points in each switching period for all the trials. The reason for fewer mesh points is that this semi-implicit integrator has a built-in Newton iteration mechanism for all its three stage equations.

Results and Analysis

First of all, some definitions are given, which will be used to quantitatively evaluate the performance of each method in this case study.

We defined the exit port average concentration within one switching as $\overline{y}$:

$$\overline{y} = \frac{y(x, t_1) + 2\sum_{i=2}^{N_t-1} y(x, t_i) + y(x, t_{N_t})}{2(N_t - 1)}, \tag{7.18}$$

where N_t is the total time steps of integration for one switching period.

For component i, the product purity Purity_i and yield Yield_i are respectively defined as

$$\text{Purity}_i = \frac{\overline{y}_i}{\sum_{j=1}^{N_{comp}} \overline{y}_j} \tag{7.19}$$

$$\text{Yield}_i = \frac{\overline{y}_i Q_{\text{exit}}}{C_i^{in} Q_{\text{Feed}}}, \tag{7.20}$$

where N_{comp} is the number of components that need to be separated.

Figures 7.6, 7.7 and 7.8 illustrate the numerically solved concentration distribution along the total columns length at the middle of the 80th switching. They are obtained under the wavelet resolution level of 4, 5, 6 and the boundary treatment degree of 1, 2, 3.

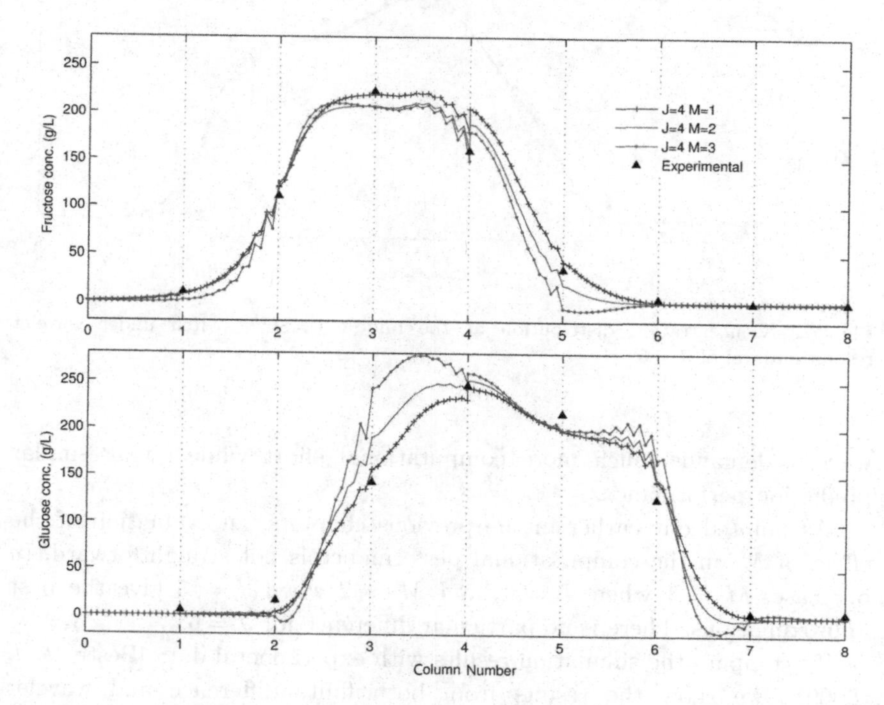

Fig. 7.6 Concentration distribution at the middle of 80th switch under wavelet resolution level of $J = 4$.

Although simulation using wavelet approach has been conducted under different levels, the results from lower levels can fit equivalently well with the experimental data, especially at the feeding port, where the concentration front experiences a sudden change. Higher level (J = 6)

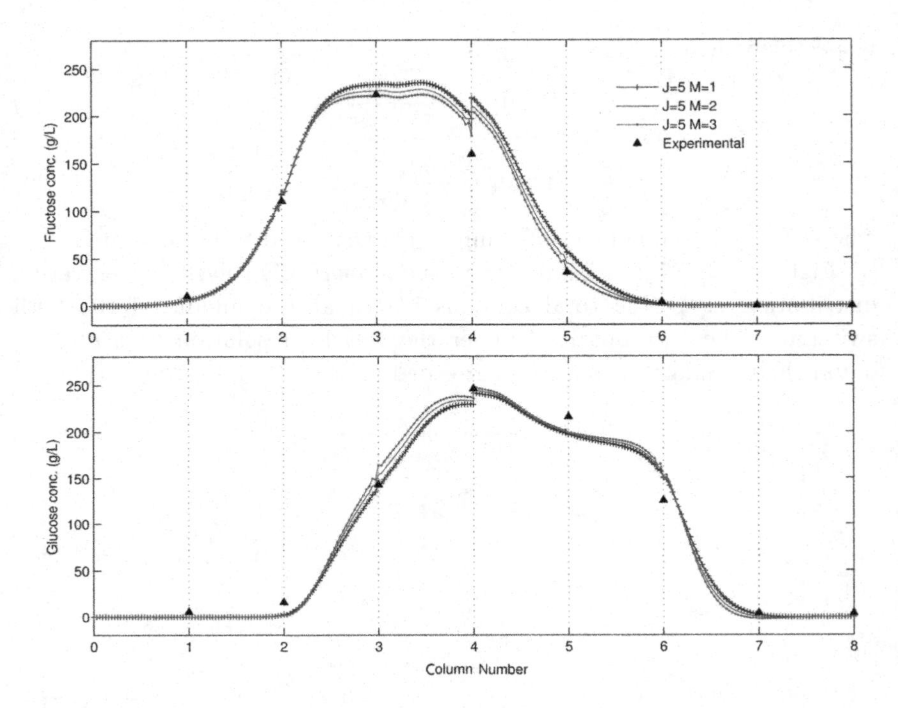

Fig. 7.7 Concentration distribution at the middle of 80th switch under wavelet resolution level of $J = 5$.

wavelet demands much more computational effort while giving similar prediction performance.

As pointed out earlier in the previous chapters, an evaluation of the effect of M on the computational performance is not straightforward. In our case, $M = 1$ when $J = 4$, and $M = 2$ when $J = 5$ give the best approximations. There is no particular difference for $J = 6$.

To compare the simulation results with experimental data [Beste *et al.* (2000)], we show the results from both finite difference and wavelet collocation methods in Figure 7.9. The two methods have generated almost identical profile. However, as far as computational cost is concerned, wavelet takes remarkably less time for each switching period (5 seconds for $J = 4$; 16 seconds for J = 5; in comparison with 285 seconds for $Nx = 65$ using the finite difference method in this study).

From the time profile of average concentration in Figure 7.10, numerical simulation from wavelet reaches steady state faster than the solution from

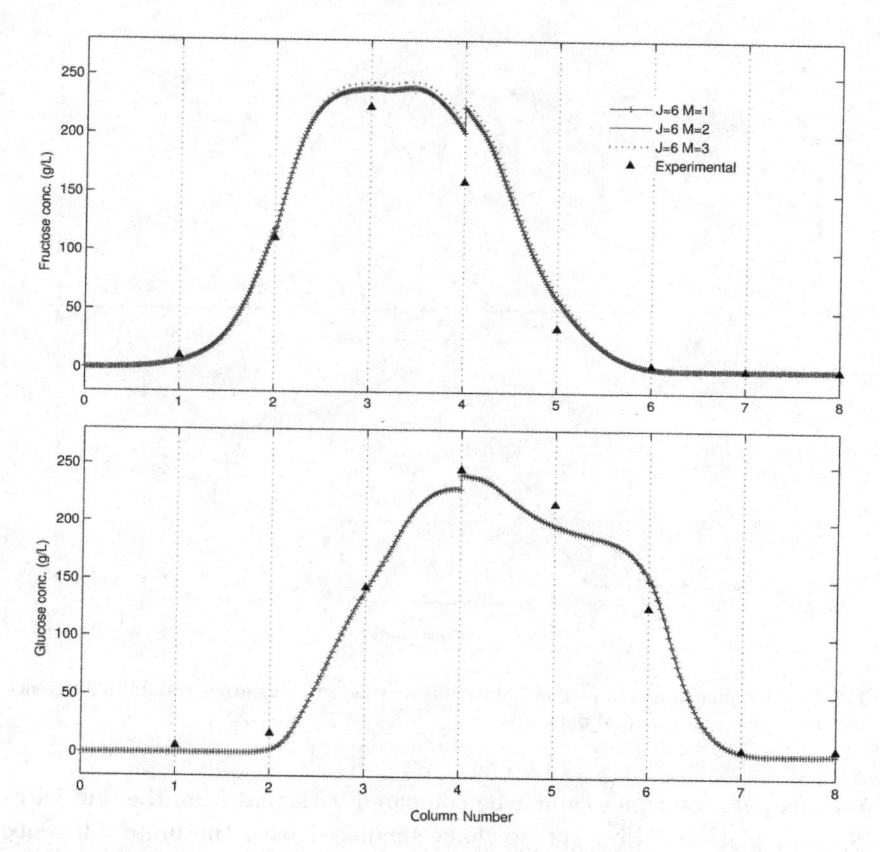

Fig. 7.8 Concentration distribution at the middle of 80th switch under wavelet resolution level of $J = 6$.

finite difference.

Furthermore, the product purity and yield are used as criteria for evaluation of separation quality. They are calculated as an average during a switching period. Table 7.2 lists the results from our simulation, reported experimental data, and other published simulation results based on the same case study. The numerical solution using wavelet is very close to the results reported in [Lim (2004)], which was carried out on Sun Ultra Spark I platform using the Method of Line.

As for the computing demand and efficiency, the reported calculation time for the Method of Line (MOL) on Sun Ultra Spark I platform is 4-5 hours. The numerical simulations of this project have been conducted on a personal computer with an Intel's Pentium IV 3.00GHz processor, and thus

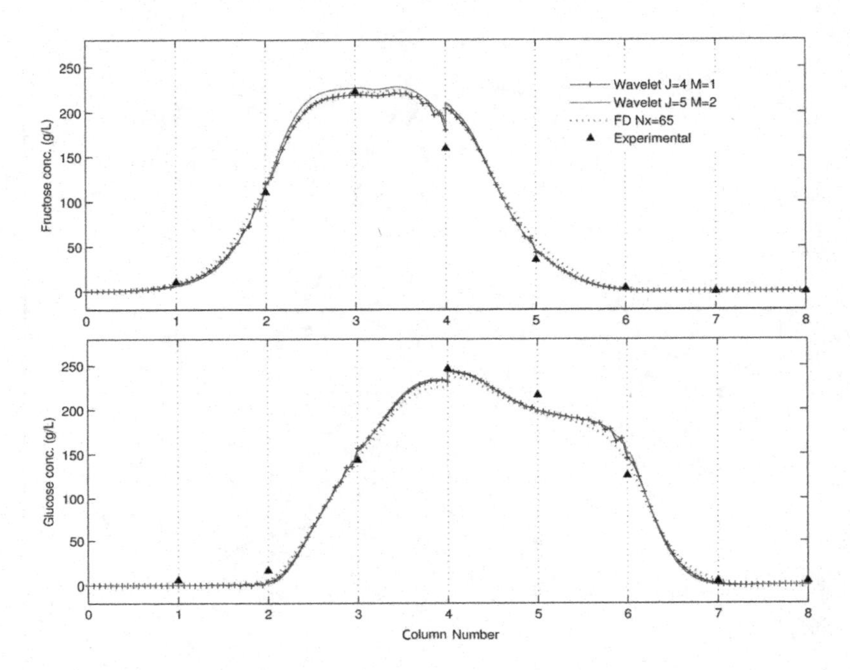

Fig. 7.9 Comparison of numerical solution from wavelet-collocation and finite difference methods with experimental data.

the computation times cannot be compared with that from the Sun Ultra Spark I platform. However, we have simulated both the finite difference method and the wavelet collocation method on the same personal computer, and therefore the computation times can be directly compared with each other.

It is seen from Table 7.2 that the finite difference method consumes a few hours to get the results, while the wavelet method requires only 6 min to 21 min. This indicates that the computing performance of the wavelet method is encouraging. The track of purity and yield at extract and raffinate ports shown in Figure 7.11 present a good convergence property for both methods.

7.5 Case Study II: Bi-naphthol Enantiomers Separation

The separation of Enantiomers of 1,1'-bi-2-naphthol in 3,5-dinitrobenzoyl phenylglycine bonded to silica gel columns reported by [Pais *et al.* (1997)] is taken as our second case study system. The SMBC process also involves

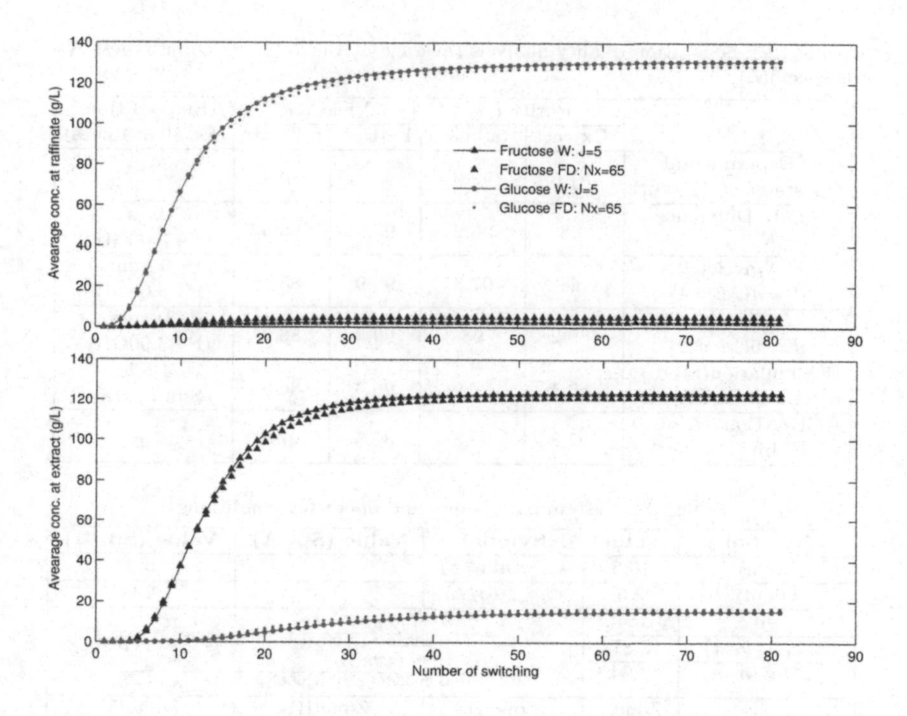

Fig. 7.10 Time profile of average concentration at the extract and raffinate ports.

8 columns with a configuration of 2:2:2:2. System parameters and operation conditions are listed in Table 7.3.

As the Peclet number is close to 2000 in this application, the finite difference method will not be adopted for solving the model equations numerically due to the steepness of the wave front. Therefore, numerical simulations have been performed using both the high resolution and wavelet collocation methods for spatial discretisation.

For the trials of the high resolution method, the number of mesh points along one column length has been chosen to be $N_z = 17$ and 33, which are equivalent to the collocation points generated by wavelet level of J = 4 and J = 5, respectively. Simulations using wavelet collocation method are conducted on the level J = 4, 5, and 6, respectively. The boundary conditions are treated using polynomial interpolation with the degree M = 1. The number of mesh points along the time axis is 5 points each switching period for all the trials.

Table 7.2 Separation quality analysis (average at the 73th switching unless N_s is specified).

	Purity (%)		Yield (%)		Elapsed time
	Extract	Raffinate	Extract	raffinate	(for 80 switches)
#Experimental ([Beste *et al.* (2000)])	81.6	92.9	96.4	80.4	-
Finite Difference $N_x = 65$	88	96.2	97.6	85.5	6 hr (P4 3.00GHz)
Wavelet $J = 4(M = 1)$	88	97.8	98.9	85.1	6 min (P4 3.00GHz)
Wavelet $J = 5(M = 2)$	89.2	98	99.3	86.9	21 min (P4 3.00GHz)
#Simulation(MOL) ([Beste *et al.* (2000)])	89.4	97.9	98.3	86.9	4-5 hr (Sun Spark I)
#CE/AE at $N_s = 40$ ([Lim (2004)])	97.8	98.3	89.5	97.3	-

Table 7.3 System parameters and operating conditions.

Symbol	**Value**	**Symbol**	**Value (Sp. A)**	**Value (Sp. B)**
L(cm)	10.5	$k_{\text{eff},i}(\text{min}^{-1})$	6.0	6.0
D(cm)	2.6	$C_{i,\text{feed}}(g/L)$	2.9	2.9
ϵ_T	0.4	$q_A^* = \frac{2.69C_A}{1+0.00336C_A+0.0466C_B} + \frac{0.1C_A}{1+C_A+3C_B}$		
$t_{switch}(min)$	2.87			
$Q_F(ml/min)$	3.64	$q_B^* = \frac{3.73C_B}{1+0.00336C_A+0.0466C_B} + \frac{0.3C_B}{1+C_A+3C_B}$		
	Zone I	Zone II	Zone III	Zone IV
$Q(ml/min)$	56.83	38.85	42.49	35.38
$D_T(cm^2/min)$	0.141	0.096	0.105	0.088

Results and Discussion

Here, we use relative error, Err, as a criterion to evaluate the convergence performance of the proposed algorithms. The metrics for relative error is defined in [Minceva *et al.* (2003)] between the average concentration of each component in the extract and raffinate stream for the successive switches.

$$\begin{aligned} \text{Err} = & \left|1 - \frac{E_{A,s-1}}{E_{A,s}}\right| + \left|1 - \frac{E_{B,s-1}}{E_{B,s}}\right| \\ & + \left|1 - \frac{R_{A,s-1}}{R_{A,s}}\right| + \left|1 - \frac{R_{B,s-1}}{R_{B,s}}\right|, \end{aligned} \tag{7.21}$$

where $E_{A,s}$ stands for average concentration of component A in the Extract stream for this switch and is defined by Equation (7.18); R is for raffinate stream.

Figures 7.12 and 7.13 illustrate the calculated concentration distributions against experimental data along the total columns length.

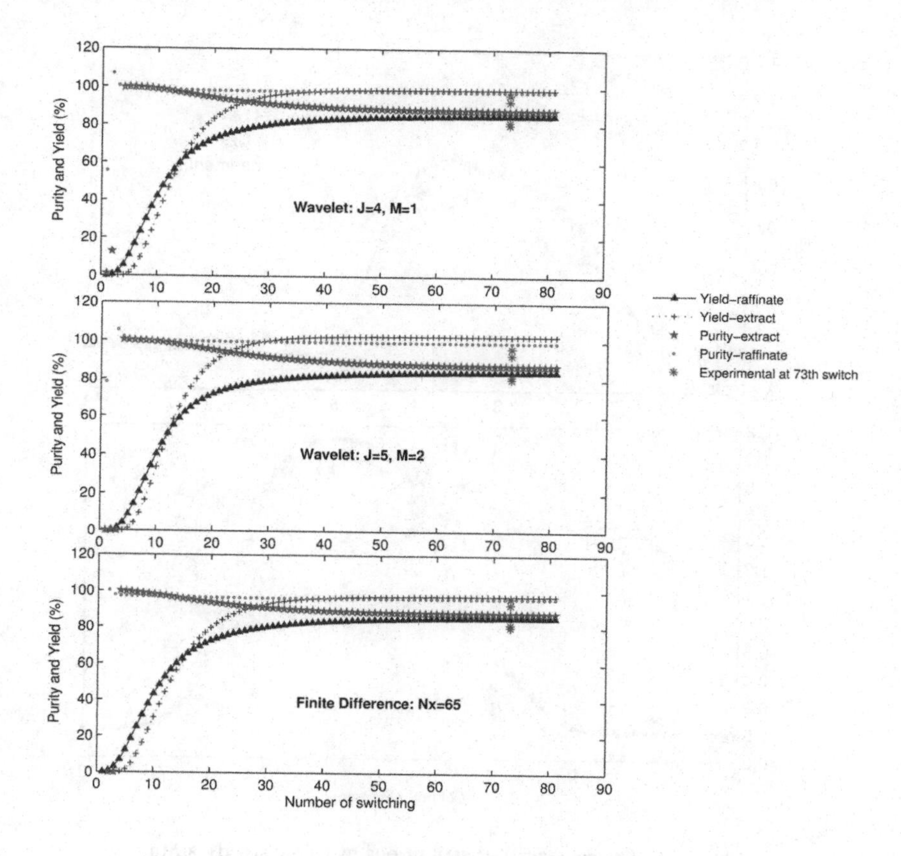

Fig. 7.11 Time profile of purity and yield at exit ports.

They are obtained at the middle of the 80th switching, which is considered to be steady state.

It is seen from Figures 7.12 and 7.13 that $J = 4$ (wavelet collocation) or $N_z = 17$ (high resolution) are not good enough to predict the real value. Furthermore, wavelet collocation presents certain degree of oscillation. However, with the increasing density of spatial mesh points, e.g., wavelet collocation with $J = 5$ or high resolution with $N_z = 33$ (Figure 7.13), the numerical approximations are getting better and the high resolution method shows much closer results. As far as computational cost is concerned, using the same number of spatial mesh points, wavelet takes less time for each switching period (16 seconds for $J = 5$) because less iteration (only 2) is required during the solving of Jacobian matrices. While for

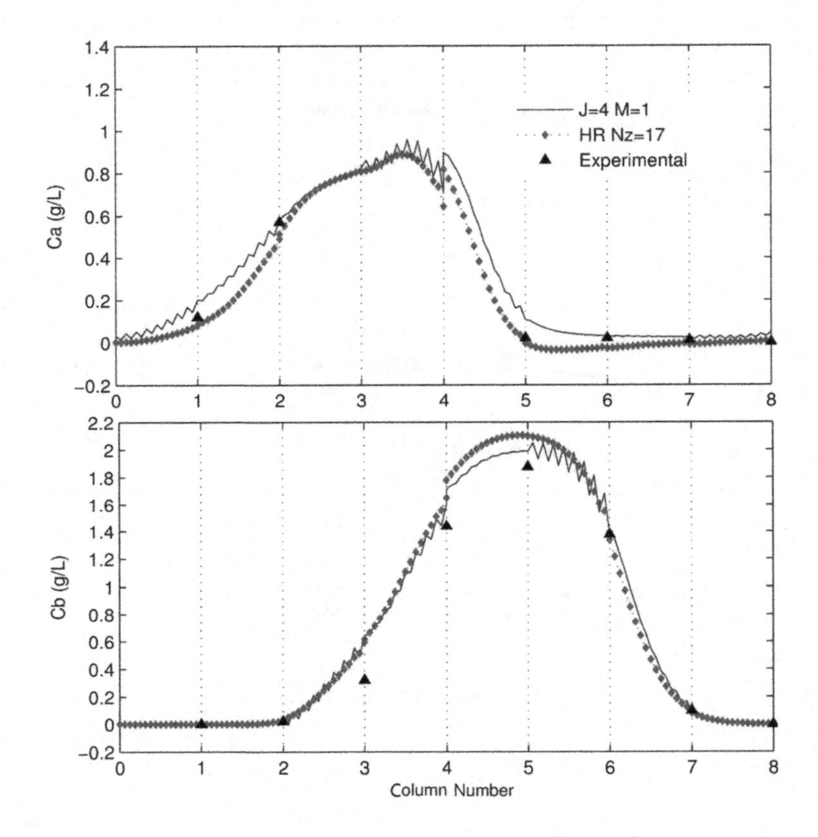

Fig. 7.12 Concentration distributions at cyclic steady state.

high resolution method, the computation consumes 24.8 seconds for one switching calculation in which 4 iterations are required. Nevertheless, the results from the high resolution method are much closer to the reported experimental data.

Further analysis is carried out by examining the relative error produced by wavelet $J = 5$ and high resolution $N_z = 33$. Figure 7.14 gives the relative error from wavelet collocation with different interpolation degrees. The figure is accomplished by two scaling down sub figures. A common observation is the abrupt point at the cycle transaction point ($s \times 8$ switch). The cause of this is not clear at the moment.

As demonstrated in Figure 7.15, high resolution also has consistent and better convergence performance.

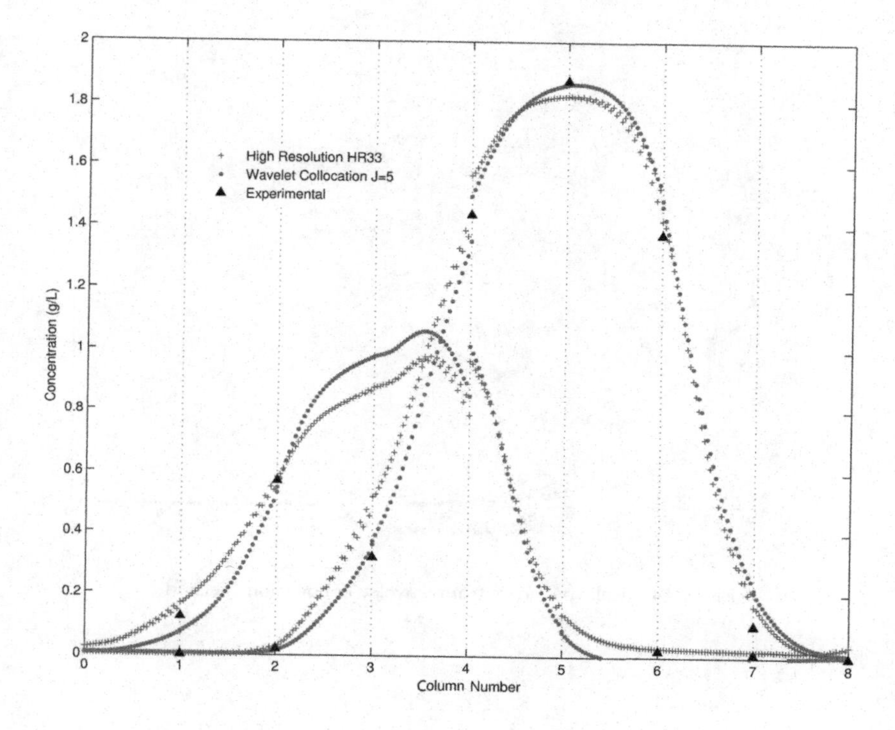

Fig. 7.13 Concentration distributions at cyclic steady state.

7.6 Concluding Remarks

This chapter has explored some upfront discretization techniques for the solution of complicated dynamic system models with sharp variations. Recently developed wavelet-based approaches and high resolution methods have been successfully used for numerically solving mathematical models of SMBC separation processes.

The investigations in this chapter show that both wavelet and high resolution methods are good candidacies for the numerical solution of this type of complex mathematical models. Both methods have generated encouraging results in terms of computation time and prediction accuracy on steep front. High resolution presents better stability at achieving steady state and closer approximation to experimental data. Wavelet-based methods have the advantage of capturing system dynamics at low resolution levels, therefore, require less computational effort. The attempts made in

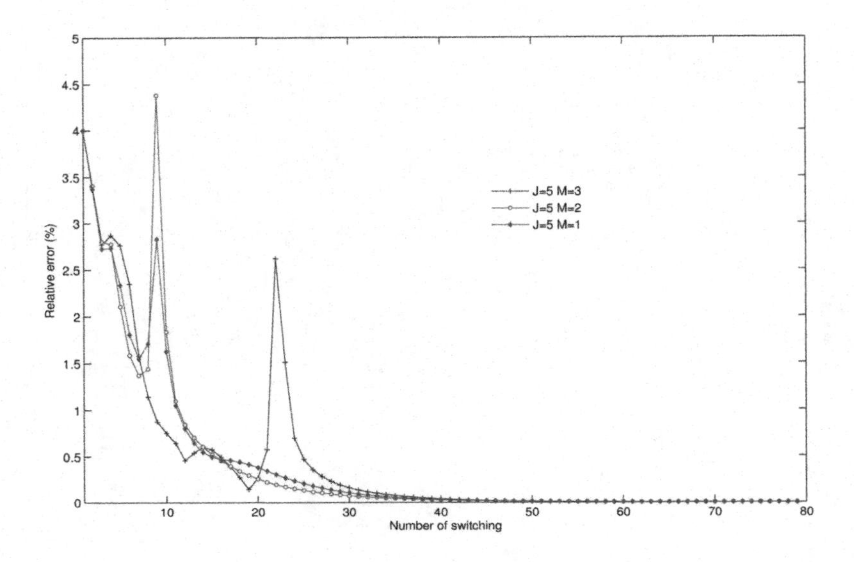

Fig. 7.14 Relative error from wavelet collocation method.

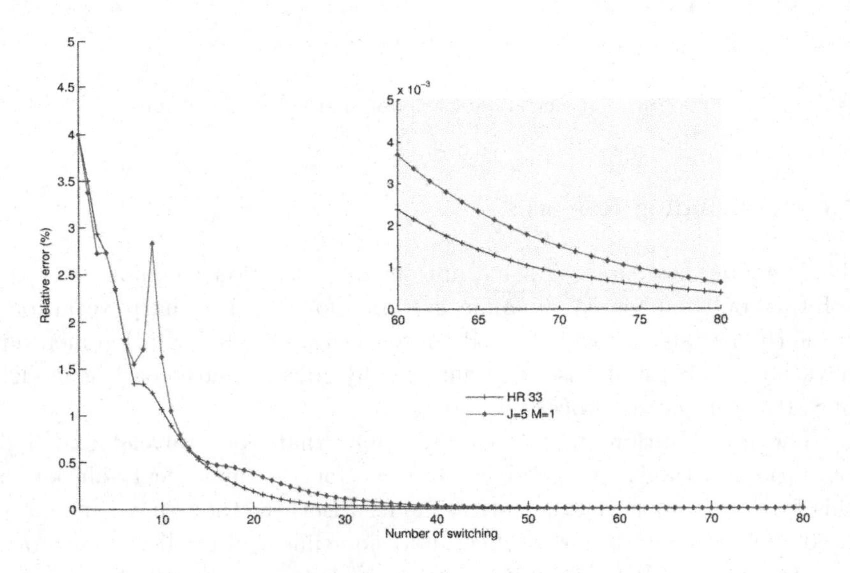

Fig. 7.15 Comparison of relative error.

this research prove that it is not always necessary to use a sophisticated mathematical tool if an algorithm with its basic structure can solve the problem.

The case studies in this chapter also provide a powerful numerical computing framework for solving complicated SMBC process models, as well as other complex industrial process models. The framework benefits on-line optimization and real-time control of complex industrial processes.

Chapter 8

Conclusion

Numerical computation of mathematical models is a very broad topic in applied mathematics. But this book has investigated numerical computation mainly from the practical application perspective, and the application areas we have focused on are complex industrial processes. From this perspective, we have treated numerical computation of various mathematical models as a process of problem solving in this book for complex industrial processes. This has motivated us to discuss not only the computational methods but also all major steps of the whole problem solving process. The steps we have covered in this book include process and requirements analysis, process modelling, development of numerical approximations, derivation of numerical solutions to the models, and interpretation and verification of the numerical results.

With problem solving in mind, the performance and accuracy of the numerical computation not only reply on the numerical methods chosen for the computation, but also depend on process modelling before the numerical computation is actually conducted. Therefore, special attention has been paid to process modelling in this book. Without a well established mathematical model that accurately describes the process dynamics, good computing results cannot be expected regardless what numerical method is chosen for the numerical computing task. Obviously, in order to establish a good process model, process analysis is crucial to clarify the requirements specifications.

We have noticed that a numerical computation task serves a specific purpose in a real-world application. In this book, we have limited our application purposes on online optimization and real-time control of complex industrial processes. In modern process industries, most industrial processes are well understood and can be well controlled by using some simple strategies, such as proportional-integral-derivative (PID) control. They do not need advanced model computation algorithms for process optimization and control. However, there are also many processes with complex dynamics and/or a large number of degrees of freedom. Also, it is a trend in modern process industries to integrate multiple conventional processes into a compact but more complicated unit operation. Mathematical models established from those processes demand significant computing resources for numerical computing. They have been addressed throughout this book.

Depending on the complexity, scale, and dynamics of the mathematical models of the processes, numerical computation of the process models have been investigated in different ways in this book. Our investigations have included the simple Runga-Kutta method, more advanced finite difference methods, wavelet-based numerical methods, and high resolution methods. Finite difference methods are easy to derive, and are suitable for large-scale model computation problems. As relatively new numerical techniques, wavelet-based methods have some useful features and are computationally efficient. The high resolution methods are particularly effective for stiff problems with steep changes in the shape of the solutions.

To facilitate our investigations into the numerical computation of mathematical models for complex industrial processes, we have studied a number of practical industrial processes in this book. The processes we have discussed include biological fermentation, continuous galvanizing, chemical reaction, chromatography and crystallization. They are not only widely deployed in process industries, but also represent typical types of significant process unit operations. Numerical techniques developed for these processes are also applicable to other industrial processes with similar dynamic behaviours. For example, numerical methods for crystallization processes are suitable for other processes whose dynamics can be described using population balance equations. In this sense, our investigations into the numerical computation for selected industrial processes in this book have important insights to many other complex industrial processes.

Discussions of numerical computation with comprehensive case studies are an important feature of this book. Particularly, we have provided step-by-step procedures to deal with these case studies with simulation and/or experimental verification. These case studies and practical examples form some useful frameworks for numerical computing of the mathematical models of complex industrial processes. Without major modifications, these frameworks can be applied to many other process systems.

This book has summarized many of our previous research outcomes in the broad area of computational theory and applications. While we have not claimed new technical contributions in this book, we have well organized our previous contributions into several categories as the chapter and section headings of the book have suggested. With this organization, we have developed a better understanding of the feasibility and suitability of numerical computing methods in specific applications. We have also developed some new insights into the applications of various numerical methods. For example, with comprehensive comparative studies, we have understood that among all numerical methods we have investigated in this book, high resolution methods are most suitable for stiff systems.

Finally, the presented work in this book on computation of mathematical models for complex industrial processes suggests some interesting research directions. We briefly describe a few of them in the following.

1) As the numerical computation is for solving problems from industrial processes, better understanding the process dynamics and requirements are always the first step to conduct. Particularly, many complex or large-scale industrial processes have not been well described for specific application scenarios, leading to emerging requirements of better modelling techniques.
2) The increasing integration of multiple unit operations into a compact one in modern process industries leads to new system design, and highly coupled systems and process models with or without delay. Such systems display coexistence of multiple steady states known as multiplicity as well as change in the number of steady states known as bifurcation. These behaviours will at best lead to low-performance operation and at worst result in explosion and disasters. Therefore, identification and computation of multiplicity and bifurcation of the systems are crucial, though challenging. Computational theory and technologies are required to tackle this challenge.
3) There are many stiff problems arising from industrial processes such as crystallization. The dynamic behaviour of such stiff problems cannot be well captured using conventional numerical computing methods. Although high resolution methods have been shown to be able to handle some stiff problems, there might be other dynamic phenomena that have not been observed using existing numerical methods. Observing such dynamic phenomena and developing computational methods to

handle them are worthy to be investigated.

4) with the rapid development of hardware and software technologies, more and more computing power and resources are becoming available. Many computing algorithms that are not suitable for some process models due to the significant time consumption may become feasible in industrial applications. Therefore, further development of existing numerical computing methods for practical applications would be promising.
5) Conventional computing tasks are implemented mainly on single core computing platforms. However, chip multiprocessor systems, which integrate multiple cores into a single CPU, proliferate in recent years. How to make efficient use of the multiple cores in the CPU to speed up the computation through innovative computational theory and algorithm design will be an interesting topic of research and development.

Obviously, this is not an exhaustive list of the research directions in the broad area of numerical computation of mathematical models for complex industrial processes. With new concepts and technologies emerging in various aspects related to the area, innovative methods will also appear. Let us look forward!

Bibliography

Alt, R. (1978). A stable one step methods with step size control for stiff systems of ordinary differential equations, *J. Comput. Appl. Math.* **4**, pp. 29–35.

Azevedo, D. C. S. and Rodrigues, A. S. (2000). Obtainment of high-fructose solutions from cashew (Anacardium occidentale) apple juice by simulated moving-bed chromatography. *Separation Science and Technology* **35(16)**, 2561-2581.

Bertoluzza, S. and Naldi, G. (1996). A wavelet collocation method for the numerical solution of partial differential equations, *Appl. Comput. Harm. Anal.* **3**, pp.1–9.

Beste, Y. A., Lisso, M., Wozny, G. and Arlt, W. (2000). Optimisation of simulated moving bed plants with low efficient stationary phases: Separation of fructose and Glucose. *Journal of Chromatography A* **868**, 169-188.

Beylkin G. (1992). On the representation of operators in bases of compactly supported wavelet, *SIAM J. Numer. Anal.* **29**, pp. 1716–1740.

Biegler, L. T., Jiang L. and Fox, V. G. (2004). Recent advances in simulation and optimal design of pressure swing adsorption systems. Separation and Purification Reviews. **33(1)**, 1-39.

Chen, M., Hwang, C. and Shih, Y. (1996). The computation of wavelet-galerkin approximation on a bounded interval, *Int. J. Numerical methods in Engineering.* **39**, pp. 2921–2944.

Cohen, A. (2003). *Numerical analysis of wavelet methods* (Elsevier, Netherlands).

Daubechies, I.(1992). *Ten lectures on wavelets* (SIAM, Philadelphia).

Duduković M. P. and Lamba H. S. (1978). Solution of moving boundary problems for gas-solid noncatalystic reactions by orthogonal collocation. *Chemical Engineering Science.* **33**, pp. 303–314.

Ebrabimi H. A. and Jamshidi E. (2001). Kinetic study od zinc oxide reduction by methane. *Trans. IChemE. Part A.* **79**, pp. 62C-70.

Ebrabimi A. A., Ebrabimi H. A. and Jamshidi E. (2008). Solving partial differential equations of gasCsolid reactions by orthogonal collocation. *Computers and Chemical Engineering.* **32**, pp. 1746C-1759.

Gu, T. (1995). *Mathematical modelling and scale up of liquid chromatography.* (New York: Springer).

Gunawan, R., Fusman, I. & Braatz, R. D. (2004). High resolution algorithms for multidimensional population balance equations. *AIChE Journal* **50**, pp. 2738–2749.

Hsiao, C. H. (2004). Haar wavelet approach to linear stiff systems, *Math. Comput. Simu.* **64**, pp. 561–567.

Jin, S. H. and Wankat, P. C. (2005). New design of simulated moving bed for ternary separations. *Industrial and Engineering Chemistry Research* **441(6)**, 1906-1913.

K. T. Klasson, M. D. Ackerson, E. C. Clausen, J. L. Gaddy (1991). Modelling Lysine and Citric Acid Production in Terms of Initial Limiting Nutrient Concentrations. *Journal of Biotechnology* **21 (3)**, 271-281.

Koren, B. (1993). A robust upwind discretization method for advection, diffusion and source terms. *CWI Report NM-R9308*, April.

Kougoulosa, E., Jonesa, A. G., Wood-Kaczmar, M. W. (2005). Estimation of crystallization kinetics for an organic fine chemical usinga modified continuous coolingmixed suspension mixed product removal (MSMPR) crystallizer. *Journal of Crystal Growth* **273**, pp. 520–528.

Lapidus, L. & Amundson, N. R. (1952). Mathematics of adsorption in beds. VI. The effect of longitudinal diffusion in ion exchange and chromatographic columns. *Journal of Physical Chemistry* **56**, 984-988.

Lim, Y., Jorgensen, S. B. (2004). A fast and accurate numerical method for solving simulated moving bed (SMB) chromatographic separation problems. *Chemical Engineering Science* **59**, 1931-1047.

Liu et al. (2000) Liu, Y., Cameron, I. T. and Bhatia S. K. (2001). The wavelet-collocation method for adsorption problems involving steep gradients. *Computers and Chemical Engineering* **25**, 1611-1619.

Liu, Y. (2001). *Application of wavelets to two classes of process engineering problems.* (PhD thesis, The Univeristy of Queensland, Australia)

Liu, Y. and Tadé, M. O. (2004). New wavelet-based adaptive method for the breakage equation, *Powder Technology,* **139**, pp. 61–68.

Lu, Y., Wei, F., Shen, B., Ren, Q. and Wu, P. (2006). Modelling, simulation of a simulated moving bed for separation of phosphatidylcholine from soybean phospholipids. *Chinese Journal of Chemical Engineering* **14(2)**, 171-177.

Minceva, M. and Rodrigues, A. (2002). Modelling and simulation of a simulated moving bed for the separation of p-xylene. *Industrial & Engineering Chemistry Research* **41(14)**, 3454-3461.

Minceva, M., Pais, L. S. and Rodrigues, A. (2003). Cyclic steady state of simulated moving bed processes for enantiomers separation. *Chemical Engineering and Processing* **42**, 93-104.

Molnar, Z., Nagy, M., Aranyi, A., Hanak, L., Argyelan, J., Pencz, I. and Szanya, T. (2005). Separation of amino acids with simulated moving bed chromatography. *Journal of Chromatography A* **1075(1-2)**, 77-86.

Murray, J. D. (2002). *Mathematical Biology I. An Introduction Third Edition.* (Springer-Verlag, Berlin Heidelberg).

Mydlarz, J. and Jones, A. G. (1989). On numerical computation of size-dependent crystal growth rates. *Computers Chem. Engng.*, **13**, pp. 959–965.

Mydlarz, J. and Jones, A. G. (1993). On the estimation of size-dependent crystal growth rate functions in MSMPR crystallizers. *The Chemical Engineering Jaurnal*, **53**, pp. 125–135.

Pais, L. S., Loureiro, J. M., Rodrigures, A. E. (1997). Separation of 1,1'-bi-2-naphthol Enantiomers by continuous chromatography in simulated moving bed. *Chemical Engineering Science* **52**, 245-257.

Pirt, S. J. (1975). *Principles of Microbe and Cell Cultivation.* (Blackwell Scientific Publications, London).

Pynnonen, B. (1998). Simulated moving bed processing: Escape from the high-cost box. *J. Chromatography A*, **827**, 143-160.

Qamar, S., Elsner, M. P., Angelov, I. A., Warnecke, G. & Seidel-Morgenstern, A. (2006). A comparative study of high resolution schemes for solving population balances in crystallization. *Computers and Chemical Engineering* **30**, pp. 1119-1131.

Ramkrishna, D. (2000). *Population Balances: Theory and Applications to Particulate Systems in Engineering.* (ACADEMIC PRESS, New York).

Randolph, A. D. & Larson, M. A. (1988). *Theory of Particulate Processes* (Academic Press, San Diego).

Schulte, M. (2001). Preparative enantio-separation by simulated moving bed chromatography. *Journal of Chromatography A* **906**, 399-416.

Tian, Y.-C., Levy, D. C & Gu, T. (2004). Implementing a process model for real-time applications. In *Proceedings of the 7th Asia-Pacific Conference on Complex Systems (Complex'2004)*, Cains, Australia, 6-10 December 2004, 287-296.

Tian, Y.-C., Hou, C.-H. & Gao, F. (2000). Mathematical modelling of a continuous galvanizing annealing furnace. *Developments in Chemical Engineering and Mineral Processing* **8 (1)**, 359-374.

Tian, Y.-C., Tadé, M. O. and Yao, H. M. (2002). Optimization in modelling, operation and control of a fermentation process. In: *Recent Developments in Optimization and Optimal Control in Chemical Engineering*, Luus R. ed., Research Signpost, 155-177.

Wang, X. and Ching, C. B. (2004). Chiral separation and modelling of the three-chiral-centre-blocker drug nadolol by simulated moving bed chromatography. *Journal of Chromatography A* **1035(2)**, 167-176.

Wongso, F., Hidajat, K. and Ray, A. K. (2004). Optimal operating mode for Enantioseparation of SB-553261 racemate based on simulated moving bed technology. *Biotechnology and Bioengineering* **87(6)**, 704-722.

Xie et al., 2005 Xie, Y., Chin, C. Y., Phelps, D. S. C., Lee, C. H., Lee, K. B., Mun, S. and Wang. L. (2005). A five-zone simulated moving bed for isolation of six sugars from biomass hydrolyzate. *2005 AIChE Annual Meeting and Fall Showcase, Conference Proceedings* pp. 1675.

Yao, H. M., Tadé, M. O. and Tian, Y.-C. (2010a). Accelerated computation of cyclic steady state for simulated-moving-bed processes. *Chemical Engineering Science* 65(5), 1694-1704.

Yao, H. M., Tian, Y.-C. and Tadé, M. O. (1998). Using wavelets for solving SMB separation process models. *Industrial and Engineering Chemistry Research*

47(15), 5585-5593.

Yao, H. M., Tian, Y.-C., Tadé, M. O. and Ang, H. M. (2001). Variations and Modelling of Oxygen Demand in Amino Acid Production. *Chemical Engineering and Processing* 40(4), 401-409.

Yao, H. M., Zhang, T., Tian, Y.-C. and Tadé, M. O. (2010b). Wavelet based approaches and high resolution methods for complex process models. In Chapter 10 of *Handbook of Computational Chemistry Research*, edited by C. T. Collett and C. D. Robson, ISBN: 978-1-60741-047-8, Nova Science Publishers.

Zhang T., Tian, Y.-C., Tadé, M. O. and Utomo, J. (2007). Comments on "The Computation of Wavelet-Galerkin Approximation on a Bounded Interval", *Int. J. Numer. Meth. Engng.* **72**, pp. 244–251.

Zhang T., Tadé M., Tian Y.-C. and Zang H. (2008). High-resolution method for numerically solving PDEs in process engineering, *Computers and Chemical Engineering* **32**, pp. 2403–2408.

Printed in the United States
By Bookmasters